MIRIAM RIVERA & CLINT KORVER
304 Genoa Drive
Redwood City, CA 94065

Industry Analysis

The Oil and Gas Industries

AF770488

November 12–13, 1992
Houston, Texas

James L. Carroll, CFA
James F. Clark
David N. Fleischer, CFA
Carol Freedenthal
Edd Grigsby
Richard G. Gross II
Michael T. Kerr
Michael L. Mayer
Thomas P. Moore, Jr., CFA
John E. Olson, CFA
Thomas A. Petrie, CFA, *Moderator*
Bernard J. Picchi, CFA
William L. Randol
Arthur L. Smith, CFA
A. Paul Taylor, Jr.
Douglas T. Terreson, CFA

Edited by Thomas A. Petrie, CFA

To obtain an AIMR Publications Catalog or to order additional copies of this publication, turn to page 154 or contact:

AIMR Publications Sales Department
P.O. Box 7947
Charlottesville, VA 22906
U.S.A.
Telephone: 804/980-3647
Fax: 804/977-0350

The Association for Investment Management and Research comprises the Institute of Chartered Financial Analysts and the Financial Analysts Federation.

© 1993, Association for Investment Management and Research

All rights reserved. No part of this publication may be reproduced, stored in a retrieval system, or transmitted, in any form or by any means, electronic, mechanical, photocopying, recording, or otherwise, without the prior written permission of the copyright holder.

This publication is designed to provide accurate and authoritative information in regard to the subject matter covered. It is sold with the understanding that the publisher is not engaged in rendering legal, accounting, or other professional service. If legal advice or other expert assistance is required, the services of a competent professional should be sought.

From a Declaration of Principles jointly adopted by a Committee of the American Bar Association and a Committee of Publishers.

ISBN 1-879087-24-3

Printed in the United States of America

7/15/93

Table of Contents

(continued on next page)

Foreword

Fluctuations in supply and demand, innovations in technology, and changing regulations have dramatically changed the oil and gas industries in recent years. After more than a decade of geopolitically induced price volatility, oil companies and their consumers are facing a period of relative stability. Such structural changes have had a profound impact on the oil industry. With oil prices stabilized, oil companies are learning to survive and compete with increased use of technology and fewer tax incentives. Structural changes are also affecting the natural gas industry, which has been plagued by imbalances in supply and demand for years. Deregulation has introduced more competition to the marketplace, and the industry is struggling to adapt. Meanwhile, environmental concerns affecting both the oil and gas industries increase the likelihood of more structural changes in the future.

The Oil and Gas Industries strives to place these industry changes in perspective for securities analysts and others in the investment management profession. It is the fourth in AIMR's Industry Analysis series of seminars and proceedings, following those addressing the transportation, financial services, and retail industries. The series was conceived by Charles D. Ellis, CFA, to provide educational material on the nuances of individual industries from the perspective of security analysis. In most cases, the specific technical information that must be the backbone of any sound industry analysis is available only through personal experience with a particular industry. This series of seminars and publications makes the fruits of that experience available to all.

Each industry seminar and publication is built around an analytical framework that identifies the key factors to consider in conducting an effective analysis of the industry and that highlights the specific interrelationships that underlie sound valuation decisions. The key topics are understanding industry basics, analyzing the internal and external factors that affect the industry, interpreting the industry numbers, and valuing the industry's securities. Several of the presentations also address the problem of selecting individual stocks. *The Oil and Gas Industries* also contains a glossary of technical terms to provide readers with the vocabulary necessary to understand the industries.

The speakers at the seminar, whose presentations this proceedings reproduces in full, are among the leading specialists in oil and gas analysis. AIMR wishes to thank them for sharing their research and practical experience and for assisting in the preparation of this proceedings. Special thanks are due to Michael T. Kerr, Carol Freedenthal, and Thomas P. Moore, Jr., CFA, who helped with the glossary. Thanks also are extended to Thomas A. Petrie, CFA, who edited the proceedings and contributed the overview that expertly sets the stage for the presentations that follow.

The speakers contributing to the seminar were: James L. Carroll, CFA, PaineWebber, Inc.; James F. Clark, First Boston Corporation; David N. Fleischer, CFA, Goldman, Sachs & Co.; Carol Freedenthal, Jofree Corporation; Edd Grigsby, Phillips Petroleum Company; Richard G. Gross II, Lehman Brothers; Michael T. Kerr, Capital Guardian Research Company; Michael L. Mayer, Wertheim Schroder & Company, Inc.; Thomas P. Moore, Jr., CFA, State Street Research & Management Company, Inc.; John E. Olson, CFA, Merrill Lynch & Company, Inc.; Thomas A. Petrie, CFA, Petrie Parkman & Company; Bernard J. Picchi, CFA, Kidder Peabody & Company, Inc.; William L. Randol, Salomon Brothers, Inc.; Arthur L. Smith, CFA, John S. Herold, Inc.; A. Paul Taylor, Jr., Anadarko Petroleum Corporation; and Douglas T. Terreson, CFA, Putnam Companies.

Dorothy C. Kelly
Assistant Vice President
Publications and Research
AIMR

Biographies of Speakers

James L. Carroll, CFA, is managing director of PaineWebber, Inc. He previously served as petroleum analyst with a bank in the Midwest and later joined the Houston-based firm of Rotan Mosle, Inc. Mr. Carroll has been designated a top-ranked analyst in investment research for Petroleum Services by Greenwich Research Survey and *Institutional Investor* magazine. He is a member of the National Association of Petroleum Investment Analysts and a member of the New York Society of Security Analysts. He received his M.B.A. from Wayne State University.

James F. Clark is vice president and equity research analyst for First Boston Corporation, covering domestic and international integrated oil companies and independent refining companies. Previously, Mr. Clark was in the Investment Banking Division of First Boston. He was named to the 1992 All-American Research Team of *Institutional Investor* in the domestic oil category. Mr. Clark holds a B.A. in economics and American Studies from Williams College and an M.B.A. from Harvard University.

David N. Fleischer, CFA, is vice president of Goldman, Sachs & Company. As a member of the Investment Research Department, he specializes in the natural gas industry. Previously, he was first vice president and senior natural gas analyst at Prudential Securities, Inc., assistant vice president at J.P. Morgan Guaranty Trust Company, and a generalist in natural resources for Sanford C. Bernstein & Company. Mr. Fleischer joined Prudential as a natural resource analyst specializing in the natural gas industry. He received his B.A. in economics from the University of Pennsylvania and his M.B.A. from the Wharton School of the University of Pennsylvania.

Carol Freedenthal is chief executive officer of Jofree Corporation. He has worked extensively in various energy and chemical business areas, specializing in natural gas marketing, regulation, and engineering. Previously, he was research engineer for Monsanto, plant manager for Kennecott Cooper, general manager and chief operating officer for Kocide Chemical Company, director of commercial development for Allied Corporation, manager of market planning and development for Superior Oil Company, and senior planning consultant for Mobil Oil Corporation's North American natural gas operations. Mr. Freedenthal received his B.S. in chemical engineering from Georgia Institute of Technology.

Edd Grigsby is manager of investor relations for Phillips Petroleum Company, where he is responsible for establishing contacts within the financial community and analyzing and reporting market trends and events to management. Previously, he was director of strategic planning in Phillips Petroleum's Marketing Division, corporate administrator of federal petroleum regulations, and vice president of business development. Mr. Grigsby received his bachelor's and master's degrees in mathematics from Pittsburg State University.

Richard G. Gross II is first vice president at Lehman Brothers, where his responsibilities as senior analyst include gas price forecasting and pipeline stocks. Previously, he was at E.F. Hutton, Underwood, Newhouse & Company, and J.P. Morgan Investment Management, covering independent producers as well as pipeline and oil services. Mr. Gross holds an M.S. in finance from the University of Illinois.

Michael T. Kerr is vice president and investment research analyst at Capital Guardian Research Company, where his responsibility is the energy industry. Previously, Mr. Kerr was exploration geophysicist for Cities Service Company, executive assistant for Occidental Oil and Gas, and financial analyst for Capital Guardian Research Company. He is a member of the American Association of Petroleum Geologists and the Society of Exploration Geophysicists. He has been selected by *Institutional Investor* as a top "buy-side" analyst. Mr. Kerr received his B.A. in geological sciences and his M.B.A. from Harvard University.

Michael L. Mayer is director at Wertheim Schroder & Company, Inc. Previously, he was at Mobil Oil in various U.S. refining and marketing positions and as manager of competitive analysis and cash forecasting. He has been recognized by *Institutional Investor* for his coverage of both international and domestic oil stocks. Mr. Mayer is a member and former board member of the National Association of Petroleum Investment Analysts. He received his B.A. in economics from Stanford University and his M.B.A. in finance from the University of California, Berkeley.

Thomas P. Moore, Jr., CFA, is senior vice president and director of equity research at State Street Research & Management Company, Inc., where he sits on the Equity Investment Committee. Previously, he was president of Brinkerhoff-Signal, Inc., president of Drilling Tools, Inc., vice president of Petrolane, Inc., and security analyst with Massachusetts Financial Services, Inc. He received his A.B. degree from Providence College and his M.B.A. from Harvard University.

John E. Olson, CFA, is first vice president of securities research at Merrill Lynch & Company, Inc., where his research responsibilities include oil and gas producers, pipelines, and distributors. He previously was vice president of equity research at Rotan Mosle, Inc., first vice president of equity research at Drexel Burnham Lambert, Inc., director of equity research at First Boston Corporation, and vice president of equity research for Goldman Sachs & Company. Mr. Olson was a founding father of the National Association of Petroleum Investment Analysts. He has been on *Institutional Investor*'s All-Star Team for natural gas pipelines, producers, and oil services companies. Mr. Olson received his B.A. from the University of Pennsylvania and his M.B.A. from the Wharton School of the University of Pennsylvania.

Thomas A. Petrie, CFA, is chairman and CEO of Petrie Parkman & Company. Previously, he was vice president, senior oil analyst, and director at Wainwright Securities in New York; petroleum research analyst for Colonial Management Associates; and managing director and senior oil analyst at First Boston Corporation. Mr. Petrie has been ranked for eight consecutive years as the number one oil analyst in the exploration/independent sector by *Institutional Investor.* He has served on the Securities and Exchange Commission Advisory Board on Oil and Gas Accounting. He received his B.S. from the United States Military Academy at West Point and his M.B.A. from Boston University.

Bernard J. Picchi, CFA, is managing director of Kidder Peabody & Company, Inc. Previously, he was director of the Stock Research Department at Salomon Brothers, Inc. Mr. Picchi's oil industry research has been recognized by a number of securities industry groups, including *Institutional Investor*, Greenwich Research Associates, and *Euromoney*. He is a member of the New York Society of Security Analysts and the Oil Analysts Group of New York, of which he is a past president. He also serves on the Board of Directors of the National Association of Petroleum Investment Analysts. Mr. Picchi holds a degree in international affairs from Georgetown University.

William L. Randol is director of stock research at Salomon Brothers, Inc. Previously, he worked for First Boston Corporation both as director of equity research and senior petroleum analyst. Mr. Randol has been a member of *Institutional Investor*'s All-American Research Team for 16 consecutive years, each year placing among the top three positions in the international oil category. Mr. Randol is a member of the American Petroleum Institute, the Independent Petroleum Association of America, the National Association of Petroleum Investment Analysts, and the Oil Analysts Group of New York. He received B.A. and B.S. degrees from Rice University and an M.B.A. from Harvard University.

Arthur L. Smith, CFA, is chairman and CEO of John S. Herold, Inc., a petroleum research and consulting firm. Previously, he was at Oppenheimer and Company, Inc., the Mergers and Acquisition Department of First Boston Corporation, and Argus Research Corporation. He has testified before congressional committees on energy matters. Mr. Smith serves on the board of directors of Parker & Parsley Petroleum Company and the New York Society of Security Analysts. He is a member of the Independent Petroleum Association of America, the National Association of Petroleum Investment Analysts, and the International Association for Energy Economics. Mr. Smith received his B.A. from Duke University and his M.B.A. in economics from New York University.

A. Paul Taylor, Jr., is vice president of corporate communications for Anadarko Petroleum Corporation, an independent oil and gas exploration and production company. Previously, he served as assistant to the vice president of corporate communications and director of investor relations at Panhandle Eastern Corporation. Mr. Taylor is a member of the National Investor Relations Institute and the National Association of Petroleum Investment Analysts, and he serves on the board of directors of the Domestic Petroleum Council. He received his B.A. degree from Northeast Louisiana University.

Douglas T. Terreson, CFA, is vice president at the Putnam Companies, where he follows the energy sector and manages the Putnam Energy Resources Trust. He previously was field engineer with Schlumberger, Ltd., and has covered the energy sector at Sun Bank Capital Management and at Kemper

Financial Services. He is a member of the National Association of Petroleum Investment Analysts and a former vice president of the Chicago Energy Analyst Society. Mr. Terreson received his B.S. in petroleum engineering from Mississippi State University and his M.B.A. from Rollins College.

The Oil and Gas Industries—An Overview

Thomas A. Petrie, CFA
Chairman and CEO
Petrie Parkman & Company

After more than a decade of generally contracting industry conditions, important portions of the petroleum sector, although still confined inside a mature business, are exhibiting signs of renewed vitality. The sustainability of any such rejuvenation depends heavily on overall macroeconomic trends, because the development, processing, and distribution of conventional hydrocarbon fuels cut across virtually all important segments of economic activity. Nevertheless, the prospect of dynamic and potentially rewarding investment opportunities in important sections of the energy sector is reinforced by the likelihood of changes in the structure of the oil and gas industries. The structural changes are the results of technical innovations as well as the evolution in environmental policy making and standards. In addition, early evidence suggests that the chronic supply imbalances of the past decade may be abating for natural gas and, possibly, for liquid petroleum fuels.

This AIMR proceedings offers a timely collection of industry and sector perspectives from a number of speakers who are applying the investment decision-making art to a truly gigantic sector of economic enterprise. For perspective on the scale of the oil and gas industries, consider that the public market valuation of the 92 largest petroleum and oil services companies on a global basis exceeds half a trillion dollars.

Taken as a whole, the presentations provide a framework for judging and comparing the prospects for a wide variety of companies engaged in the petroleum, natural gas, and oil field services industries. The overall cyclical nature and maturity of the energy industries are themes that recur in a number of the presentations. As Terreson and Clark point out, the refining and marketing sector and petrochemical operations involve distinct oscillations closely tied to macroeconomic activity. In the upstream sector, the economic cycles for both crude oil and natural gas involve a long-term variability that results from complex and interrelated factors enumerated by Randol and Picchi. Some of those factors are OPEC strategy, changing environmental standards, overall growth patterns in emerging countries, and the march of Islamic fundamentalism.

In addition to the cyclical pattern of the oil and gas industries, an intriguing inverse correlation exists between monthly oil prices and U.S. Treasury bonds, as shown in **Figure 1**. The relationship suggests that, if indeed a decade-long trend toward lower interest rates is reversing, energy-hedged, yield-oriented instruments may emerge as a leading financing mechanism, especially if inflation resumes. The relationship of oil prices to T-bonds is also a key ingredient in the evolving link between energy equity securities and the property markets for underlying reserves and other associated assets.

The Clinton administration's proposal to make a tax based on British thermal units a centerpiece of its deficit-reduction plan raised numerous issues for the industries, in particular because of the different effects such a tax would have on different industry segments. Issues discussed by speakers in this proceedings run the gamut: Some speakers voice a favorable bias toward natural gas, and others warn of the challenges involved in creating different rates of tax recovery in final product prices for various liquid fuels.

The 14 presentations composing the proceedings provide insight on how energy analysts approach the various segments of the industry and are organized into two groups. The first group focuses on aspects of the integrated petroleum sector, including Randol's and Picchi's assessments of the historical evolution of the industry and reviews of factors affecting today's dynamics. This group also includes assessments of the basic characteristics of refining and petrochemicals by Terreson and Clark. Moore outlines each of the principal methodologies now in use to value companies in the integrated oil sector and provides numerous practical judgments about the methods' relative utility. Mayer's presentation on interpreting the numbers identifies several potential pitfalls that can result from blind "number crunching" in an industry that is replete with data.

The next set of presentations addresses the distinguishing characteristics of the oil field services sector. While this group is often closely linked to the other oil industry components, Carroll demonstrates that the group's underlying character is distinctly different. In fact, the link consists primarily of the fact that the rest of the industry serves as the market for oil field services. Thus, major shifts in economic patterns for energy producers, refiners, and even

Figure 1. Oil Prices versus Treasury Bonds, 1970–93[a]

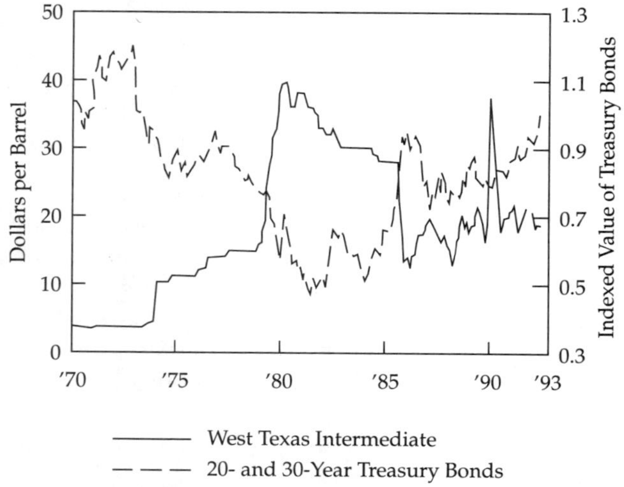

Sources: Treasury index from the Federal Reserve; oil prices from *Oil & Gas Journal.*

[a]Estimate.

distributors have important second-order implications for investment in securities of oil field services companies.

The final presentation in this group addresses the art of the industry analyst's interview. It presents useful guidelines in the context of the oil industry for improving the effectiveness of one's interviewing techniques, and it closes with a mock interview between Smith, an accomplished interviewer, and Grigsby, an experienced interviewee. This presentation has relevance to all segments of the industry, indeed to the investment analysis profession in general.

The second group of presentations in this group involves an extensive set of assessments of the natural gas industry. The considerable attention devoted to this segment is well justified by emerging industry patterns, trends in government policy, environmental concerns, and evidence of improved prospects for the supply elasticity of this commodity. Taylor develops an insightful perspective on the natural gas industry's basics by detailing its historical evolution under the myriad price-control mechanisms that have for decades distorted natural gas economics in the United States. Several speakers highlight the industry's prospects: Fleischer focuses on the factors that affect the industry's future, and Olson examines the reality that the gas pipeline sector remains a regulated business in which rate-base concepts still count. Freedenthal presents a perspective on natural gas pricing that recognizes a likely continuing competition between coal and natural gas for the foreseeable future. Finally, Gross and Kerr discuss in considerable detail various useful methods for valuing pipeline and local distribution companies.

The Integrated International Oil Companies—Understanding the Basics

Bernard J. Picchi, CFA
Managing Director
Kidder Peabody & Company, Inc.

The oil industry has evolved during three distinct eras: a "gilded" age, an age of resource nationalism, and today's age of global reintegration. Significant factors currently affecting the industry include resource distribution patterns, world politics, and the environmental movement.

The oil business produces significant revenue, income, cash flow, and employment on a worldwide scale. Revenues are about $1 trillion a year, twice the size of the GNP of Canada and roughly equal to that of Britain or Italy. Annual income, based on average after-tax profitability of $3 a barrel and a total industry production of 67 million barrels a day, is conservatively estimated at $75 billion, not including earnings from natural gas, chemicals, and other businesses. Cash flow is an estimated $200 billion a year, and worldwide employment is approximately 5 million.

History and Personalities

Oil was known to exist through natural seepage as far back as biblical times. Modern-day Iraq is the site of the first recorded seepage of oil from the ground. Until 1859, however, no one could depend on a regular supply of oil, and the principal fuels for home lighting and heating were animal fat, tallow, and whale oil.

The oil industry has gone through three significant eras: the "gilded" age, the age of resource nationalism, and the age of global reintegration.

The Gilded Age

The modern oil industry dates from 1859, when "Colonel" Edwin Drake drilled the first rotary-bored oil well in Titusville, Pennsylvania. The discovery of oil and, more importantly, how it could be recovered from underground formations revolutionized the nature of energy supply. Within 20 years, rock oil, as petroleum was then called, virtually eliminated the whale oil industry.

Until about 1915, petroleum was sought primarily for kerosene, a cheap and reliable source of home illumination. Gasoline was not in great demand until Henry Ford pioneered a way of manufacturing automobiles cheaply and making them accessible to the average citizen.

The period from the middle to the late 19th century was one of chaos in the oil fields. In 1859, northwestern Pennsylvania was the scene of perhaps the greatest land speculation ever experienced in U.S. history. People who wanted to find their fortunes headed for Pennsylvania to bid on land. Parcels of farmland that sold for $500 prior to the discovery of oil in Titusville went for $1 million only two years later. By 1865, the area had about 3,000 independent producers.

Resource control and conservation were unknown. The pressure to produce the oil as quickly as possible was tremendous; the object was to produce the oil before the neighbor drilling 100 yards away could tap the same underground deposit. As a result, oil prices collapsed during the period from 1865 to 1870.

In 1870, John D. Rockefeller decided that the industry needed organization and that the way to organize it was to control the industry's refining assets. He established Standard Oil Company and began acquiring the refining assets in northwestern Pennsylvania and northeastern Ohio. By 1882, Standard Oil, through its "trust" arrangements, controlled about 90 percent of the country's refining capacity. The trust secretly purchased refineries, storage facilities, and perhaps most important, the railroad transportation necessary to get the finished products to market. By demanding (and receiving) secret tariff rebates from the railroads, the trust was able to secure a privileged position for Standard Oil, which it used to drive competition into its arms—or into bankruptcy. Standard Oil was then in a position to deal with the several thousand unruly independent producers on its own terms.

A similar situation occurred outside the United States with the founding of Shell Oil in 1892. The founders—the Samuel brothers and, later, Detterding, the Dutch partner—applied the same organizational principles to the oil industry in the Eastern Hemisphere that Rockefeller had used in North America.

A new chapter opened in 1901 with the discovery of oil at Spindletop, Texas. Until then, Pennsylvania's oil producers were skeptical about finding any oil that was far removed from the fields of Pennsylvania and northeastern Ohio. (In fact, Henry Flagler, the top aide to Rockefeller, said he would drink every gallon of oil found west of the Mississippi; he was sure none was there.) Then, about five years after Spindletop, oil was discovered in California.

The Standard Oil trust ended in 1911 with the Supreme Court ruling in *U.S. v. Standard Oil Co.* The Supreme Court ruled that the trust violated U.S. antimonopoly laws, and the decision was later reinforced by passage of other antitrust legislation. The Court ordered the dissolution of the trust, and for the first time, real competition was introduced into the U.S. oil industry.

Even in this more competitive environment, however, the spun-out companies of Standard Oil—the ancestors of today's Mobil Corporation, Amoco Corporation, Chevron Corporation, USX-Marathon Group, and BP America—controlled the price and distribution of oil. This situation continued until the late 1960s.

The next major milestone in the industry came in 1938 with the discovery of oil in Saudi Arabia by Chevron (Standard Oil Company of California) and in Kuwait by the predecessors of British Petroleum. Even though these discoveries occurred during one of the worst years of the world depression, the companies realized that finding such a vast amount of oil would clearly change the geopolitics of petroleum.

For example, Chevron quickly recognized that it needed distribution for the enormous discovery in Saudi Arabia, because at the time, the company had no distribution facilities in the Eastern Hemisphere. This deficiency led to the formation of Caltex in the early 1940s. Caltex, a 50/50 joint venture between Chevron and Texaco, established a significant number of refineries and service stations throughout Europe and the Eastern Hemisphere.

The Age of Resource Nationalism

The world's seven largest oil companies controlled the world oil industry without interference until the Suez crisis in 1956, the beginning of the age of resource nationalism. Until 1956, the Suez Canal had been operated as a protectorate of Britain, much like the Panama Canal is now operated as a protectorate of the United States. In 1956, Gamal Abdel Nasser, the fiery and charismatic president of Egypt who deposed King Farouk, seized the Suez Canal and proclaimed that Egyptians—or, more important, Arabs—should control their own economic destiny. The seizure of the canal inspired the Arab world to take greater control of the ownership, distribution, and pricing of oil, their most important natural resource.

OPEC was founded in 1960, but the world did not recognize OPEC's importance until the Arab oil embargo of 1973. At that time, Israel was at war with Egypt and Syria and was not doing as well as it had during the Six Day War of 1967. Over protests from the Arab world, including Saudi Arabia and Kuwait, the United States and Western powers airlifted much-needed military supplies and materials to the Israelis, which changed the course of the war and allowed Israel to achieve victory. In reaction to this intervention, Arab oil producers launched the oil embargo, the culmination of its rising resource nationalism. Oil prices exploded from $3 a barrel to $25 a barrel.

Another upset occurred in 1979 during the Iranian revolution in which the Shah of Iran was deposed by the Ayatollah Khomeini. All foreign oil companies were expelled, and Iranian oil production went from 6 million barrels a day to virtually nothing for six months. The result was the second largest oil price spike in the modern era; oil prices jumped from $15 a barrel to $40 a barrel.

This era was defined by influential personalities and resource owners who were no longer content to sell their oil with the 50/50 revenue split of the past. Although his reign lasted only about six months, Iran's Mohammed Mossadegh, for example, was as much a personal precursor of events to come as was the Suez Canal's seizure a precursor of a shift in thinking among resource owners. The Ayatollah Khomeini upset world stability when he successfully gained control over Iran's oil supply. Enrico Mattei, the 1965 founder of the first modern multinational state-controlled energy company, Ente Nazionale Idrocarburi (ENI), performed a valuable service for Middle Eastern resource owners by acting as a new competitor for the seven major oil companies that cozily shared the world's greatest resource wealth in the Middle East. Mattei argued that Iranians could get better terms for the sale of their oil by dealing with his ENI group than by dealing with British Petroleum or any other of the Seven Sisters. His success encouraged upstarts such as Armand Hammer of tiny Occidental Petroleum, which negotiated deals in Libya with Muammar Qaddafi and others similar to the deals Mattei had arranged in Iran. Others who aided resource nationalists were J. Paul Getty (Getty Oil), Leon Hess (Amerada Hess Corporation), Robert O. Anderson (Atlantic Richfield), and the Donnell family (Marathon Oil).

The Age of Global Reintegration

The 1979 oil shock was followed by a period of sharply reduced oil demand and consequent overproduction by OPEC producers. The resulting oil price collapse in 1986 marked the beginning of the present age of global reintegration. The resource nationalists realized that, without ownership and control of distribution networks, they faced the same problems that beset the independent oil producers in the 1870s. There has been a drive, particularly among the major resource owners—Saudi Arabia, Kuwait, Venezuela, and Mexico—to reintegrate to combine their ownership of resources with control of the refining, transportation, and marketing assets that remain in the hands of consuming countries and the large private-sector oil firms. This reintegration may bring about a period of relative price stability and consumption growth in the oil industry.

Resource Distribution

In 1970, geologists estimated that the world had about 600 billion barrels of oil. **Figure 1** shows how oil was distributed among Latin America, Europe, Asia, Middle East, and North America in 1970 and 1992. In 1970, the Middle East, with roughly half of the world's known reserves, was recognized as the dominant player in resource distribution. North America was estimated to have 50 billion barrels, or about 8 percent of the world's oil reserves.

The absolute pie has increased, and so has the relative importance of the Middle East. In 1992, oil reserves were an estimated 1 trillion barrels, more than 50 percent greater than thought 22 years earlier. The Middle East has 660 billion barrels, or roughly 66 percent of the world's known reserves. North American oil reserves have diminished, both absolutely and proportionately, to an estimated 40 billion barrels, or about 5 percent of the total.

The total quantity of known oil reserves is a dynamic number. Exploration and drilling around the world are constantly adding to the quantity of oil reserves. Even in well-explored areas, improvements in oil recovery techniques add to reserves. When Alaskan oil was discovered in Prudhoe Bay in 1968, British Petroleum and Arco Chemical, the field's operators, estimated the field's total recoverable oil at 9–10 billion barrels; they assumed a recovery rate of 39–40 percent. Today, 15 years after Prudhoe began operation, the operators estimate that ultimate reserves will be about 12 billion barrels. The increase is attributable entirely to improvements in the technology of oil production. Technologies that were either embryonic or unknown only 10 years ago are in common commercial use today, which is how "resource banks" such as Prudhoe Bay continue to grow their reserves.

Reserve-to-production ratios vary significantly throughout the world, as **Figure 2** shows. Middle

Figure 1. Distribution of Oil Resources

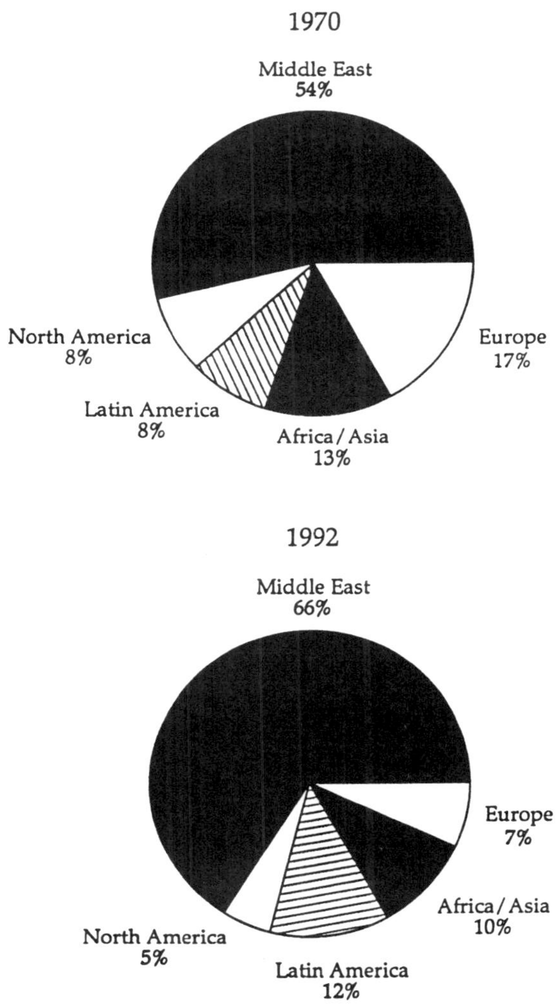

Source: BP Statistical Review of World Energy, 1992.

Figure 2. Use of Crude Oil Reserves by Major Producing Area, 1991
(reserve-to-production ratios in years)

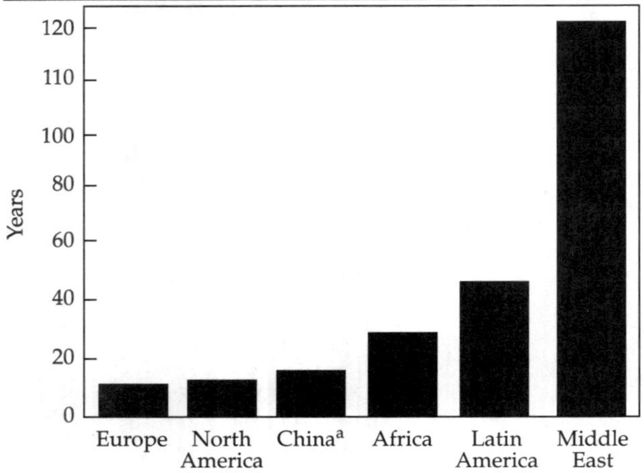

[a]Includes Commonwealth of Independent States.

Source: BP Statistical Review of World Energy, 1992.

Eastern producers have a reserve ratio of 120 years, which is more than twice the ratio of any other oil-producing area. European producers, on the other hand, with only 14–15 billion barrels in reserve and a reserve life of slightly less than 10 years, work their resource base the hardest. Over time, Middle Eastern producers will probably also trend toward lower ratios by working their resources harder.

OPEC has failed to maintain its share or even its absolute level of production since the first oil shock of 1973, which is why many OPEC countries are interested in reintegration with refining and distribution: Reintegration would provide a captive market for their ever-rising production. In 1972, the year before the Arab oil embargo, total world oil production was 58 million barrels a day. The average price of oil for the year was $2.83 a barrel. The 13 countries that composed OPEC then produced 31 million barrels, or 53 percent of total world supplies. Almost two decades later, the world price of oil is $17.50 a barrel, but the distribution of oil production has changed dramatically. Non-OPEC producers, who previously had 47 percent of the world's oil supplies, now provide nearly 66 percent.

Why has this change occurred? The tremendous increase in prices after 1972 was a major incentive for oil companies in the non-OPEC world to explore for new oil and also to develop and produce previously discovered oil that was not economically developable when oil prices were less than $3 a barrel. The greatest burst of activity in non-OPEC oil development occurred between 1973 and 1983. Norway and the United Kingdom discovered and developed major oil resources. The former Soviet Union became a significant oil exporter with the discovery of vast quantities of oil in western Siberia. Mexico became a much bigger producer than previously after it discovered oil in the Bay of Campeche, and small countries such as Malaysia, India, and Thailand became oil producers for the first time.

Politics and Economics

Politics and economics define the relationships among host countries, resource owners, oil-producing operators, and the private-sector oil extractors who operate in the host countries. In oil-producing countries, governments are usually partners of the producing entities. Oil is simply too lucrative for countries not to tax or confiscate. Currently, West Texas Intermediate, the benchmark against which most other oil prices are compared, trades at $20 a barrel. In most parts of the world, the actual direct cash cost of producing oil is only $1–$2 a barrel. That differential between price and operation cost is a juicy fiscal target for oil-owning countries to tax.

For many oil-producing countries, petroleum is an excellent and obvious instrument of economic policy. This connection is clear among Middle Eastern oil producers such as Saudi Arabia and Kuwait, but even in Britain, oil has been an important component of economic policy and growth since 1972. Some academic economists estimate that, during the past 15 years, the development of oil in the North Sea added a half of a percentage point each year to Britain's GNP growth. Because Britain did not enjoy much real economic growth during that time, this contribution had a significant impact.

Oil taxes have also been used as an instrument of demand policy. In the major economies of Europe—Germany, France, Italy, and the United Kingdom—oil products are heavily taxed at the consumer level. The average tax on gasoline in those countries today is about $3 a gallon, or $120 a barrel. Some argue that such high gasoline taxes are justified as part of environmental policy as well as economic policy, but despite the high taxes, oil consumption continues to increase in Europe.

Current fiscal systems related to oil exploration and drilling vary according to location and traditional practices in the area. These arrangements include the following:

- *Bonus and flat royalty.* This system is used in both the United States and Canada. Producers pay a bonus to landowners, either private or public, to obtain the right to drill and produce oil on a lease. If oil is found, production secures the lease for the producer, and the producer pays a flat royalty plus income tax. In Canada, this system is sensitive to the economics of individual wells, and royalties are scaled. In the United States, particularly on government lands, the fiscal system is fairly flat and economically insensitive.

This system is simple to understand and administer, but it is not very sensitive to changes in prices or the marginal economics of producing oil. For example, it does not differentiate between the costs and risks associated with drilling wells and producing oil in the Gulf of Mexico from water depths of 5,000 feet and the dramatically lower costs and risks of producing oil from water depths of 50–100 feet.

- *Bonus and progressive tax.* The United Kingdom and Norway use this system, which can be tailored to make it sensitive to costs, risks, and varying development time tables. For example, when oil prices collapsed in 1986, oil exploration within Britain was in danger of being wiped out. The government introduced important changes in its fiscal system, however, including significant tax breaks, to encourage companies to produce oil found in difficult-to-develop areas. Recently, the U.K. government proposed eliminating tax breaks on exploration, but the government will need to reintroduce some exploration tax incentives eventually, because the impact on exploration will be too severe without changes.

- *Production-sharing contracts.* The production-sharing contract (PSC) was pioneered in about 1962. It is particularly common in such countries as Indonesia and Nigeria, which do not have the ability to fund their share of capital development. In the production-sharing contract, the operator incurs all development costs and is allowed to recover 100 percent of such costs plus an agreed return out of initial production. Production revenues are then usually split 85/15 between the government and the operator. Initially, a PSC produces high rates of return for the operator but no tax revenues for the government and no tax benefits from the government. The Russians are trying to implement this system with Elf Aquitaine, the French oil company with which they have a comprehensive agreement for countrywide exploration. Elf Aquitaine could spend billions of dollars in Russia while risking that a new Russian government will repudiate its predecessor's contracts.

- *Joint ventures.* Development of the big discoveries in Colombia and, possibly, the Commonwealth of Independent States will be through joint ventures in which the governments are equal partners with the private sector almost from the start. Whether this system will be widely used elsewhere is unclear. Although joint ventures are based on 50/50 ownership, governments usually "back into" their 50 percent share of the oil resources only after the initial, high-risk exploration work has been completed by the operator.

- *Contract or fee business.* Some OPEC producers, such as Abu Dhabi, Iran, and Kuwait, use the contract, or flat-fee, arrangement to produce oil. They invite Western oil companies into their countries to produce oil for a fixed fee rather than on an equity basis. The fees are typically 50 cents to $1 a barrel and are completely insensitive to the price of oil; the fee remains the same whether oil is $40 a barrel or $4 a barrel. Operating companies can gain exposure to areas of strategic importance through this system, but equity rewards are absent, and it has generally been eschewed by Western firms except state-controlled entities such as Elf and ENI.

- *Government ownership.* Full government ownership and control of oil is the most common fiscal system in place in the world today. Most of the 8 million barrels a day of production in the former Soviet Union are wholly government owned, as is the oil production of China, Mexico, and most of the OPEC countries. This system offers no benefits to consumers, and it is demonstrably inefficient. Governments all over the world are thus loosening their grip on natural resources for economic reasons. They cannot compete in today's world without introducing competition to their own resource sectors.

Industry Segments

The modern petroleum industry is divided into four segments: upstream (oil and gas production activities), midstream (transportation, field processing, and gas gathering), downstream (refining, product marketing, and distribution), and chemicals/minerals. **Table 1** provides an estimate of how the 1991 income of the 20 largest publicly owned oil companies in the world was distributed among these categories.

Note that upstream activities accounted for the majority of the income—$12.9 billion, or 63 percent—

Table 1. Income of 20 Large Oil Firms by Activity, 1991

Activity	Income ($billions)	Percent
Upstream	$12.9	63%
Midstream	3.0	15
Downstream	7.6	37
Chemicals	2.1	10
Minerals	0.1	—
Corporate	(5.3)	(25)
Total	$20.4	100%

Source: Kidder Peabody & Co., Inc., from shareholder reports of the largest private-sector oil firms.

of these firms' total earnings of $20 billion in 1991. Downstream products, however, which include consumer nondurable items sold in convenience stores, provide significant income. The chemicals and minerals categories include value-added petrochemicals produced from natural gas liquids, naphtha, and aromatics, as well as minerals such as coal and copper.

Table 2 shows various profitability measures for typical downstream and upstream operations. The downstream business is typically low margin. With gross revenues, including taxes, of $40 a barrel in the United States, net pretax income is about $1.30 a barrel. Since 1987, the top 23 companies in the downstream business have experienced after-tax earnings of only about 80 cents a barrel. Corporate overhead deducts another 25–50 cents, which produces an estimated rate of return of 5–7 percent on invested capital. The upstream business, with a 12–15 percent return, is much more remunerative, but the return is calculated on the basis of marginal, not historical average, returns.

Table 2. "Typical" Business-Segment Profitability
(dollars per barrel, except as noted)

	Downstream	Upstream
Revenues	$40.00	$20.00
Expenses		
Excise taxes	8.00	2.00
Crude costs	20.00	—
Lifting/other	8.00	7.00
Capital recovery	2.00	3.00
Pretax	2.00	8.00
Income taxes	0.70	2.75
Net income	1.30	5.25
Return on investment	5–7%	12–15%

Source: Kidder Peabody & Co., Inc., estimates.

Price Mechanisms and Determinants

Oil prices have historically been characterized by a few very volatile periods and long periods of price stability. Between 1859 and 1880, oil prices were set by haggling, and the negotiation process was informed by a thriving futures market in crude oil. **Figure 3** shows oil prices from 1861 to 1992.

The formation of the Standard Oil trust in 1882 led to a period of relative stability in real oil prices after 20 years of extreme price volatility. In 1880, one of the earliest actions taken by Standard Oil was the purchase and disbanding of the oil exchange in New York. The price of oil transactions became invisible, to outsiders, which enhanced the trust's ability to set and control prices. At the same time, prices were set overseas by the Nobel family. (Although most people today associate Nobel with the discovery of dynamite, the Nobel family, assisted by Europe's Rothschild family, made its fortune in the discovery and exploitation of Russian oil in modern-day Azerbaijan.)

Modest pricing instability followed the termination of the trust in 1911; oil prices were set by contracts between refiners and crude suppliers, which were frequently one and the same. The period during World War I was volatile but was followed by a long period of price stability. Real oil prices were stable from 1930 to 1973.

Oil prices are ultimately set by supply and demand, but politics have had a pronounced impact on oil prices over short periods (e.g., the Persian Gulf crisis of 1990–91). The premises of the oil nationalists and the OPEC producers in the early 1970s were that (1) their oil was worth more than the prevailing price, (2) the oil producers were entitled to much higher portions of the revenue than they were receiving, and (3) prices and shares of revenue could be set unilaterally. Between 1973 and 1980, these premises guided the actions of oil producers, and the results were disastrous. This era of resource nationalism produced great price disturbances that have only recently settled down to a low volatility similar to that of the early years of the industry.

The modern period of global reintegration began in 1986. With the collapse of oil prices, Middle Eastern producers moved to integrate their oil production with the suddenly much more profitable refining activities. In addition, the oil exchanges, such as the New York Mercantile Exchange, have been resurrected since 1980. By making oil transactions transparent, understandable, and accessible to all, the exchanges have probably done more than any other entity to "democratize" oil, stabilize prices, and encourage growth in consumption by allowing consumers to manage risk.

An important determinant of the price of oil is finding cost—the actual cost spent to find a barrel of oil. During the late 1970s, finding costs were much

Figure 3. Real Oil Prices, 1861–1992
(1992 dollars per barrel)

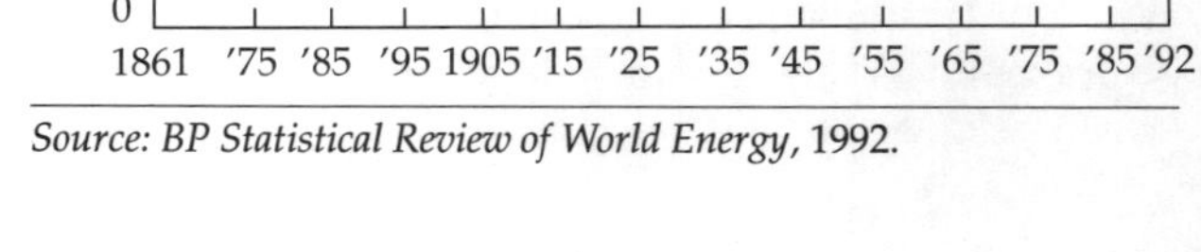

Source: BP Statistical Review of World Energy, 1992.

higher than they are today. For example, Arco and British Petroleum have improved the recovery rate of Prudhoe Bay oil from 40 to 50 percent. That boost reduced the finding costs associated with the original discovery. When oil finding costs first began to turn down in the early 1980s, many analysts said the reason was "high grading" of prospects by the industry—that is, turning away from the riskiest explorations and searching only in secure areas where the producers know oil exists.

Something more powerful than high grading is at work, however. A technological revolution has taken place in oil discovery as real-time measurement techniques, which were pioneered only in the past 15 years, have won widespread acceptance. Measurement while drilling (MWD), for example, allows operators to use the mud pulses they receive from the head of the bore hole, which could be thousands of feet below the surface, to make highly successful judgments about the formations through which they are drilling while they are drilling. To obtain that information before MWD, the operator had to take the drill string out of the well and run an "open hole" logging device into the well base far below the surface. This process was enormously complex, time-consuming, and expensive.

Over the long term, the ratio of oil prices to finding costs has been about three to one. Today, the ratio is about five to one. This increase does not mean oil prices will collapse in the near future, but costs are not pushing oil prices upward. Until monetary policies in Washington, Tokyo, or the capitals of Europe change, oil prices are not in danger of rising sharply.

Environmentalism and Energy

The environmental movement presents some great challenges to the world oil industry, especially in the United States, where growth in demand is small and resources are diminishing. The specific challenges of the environmental movement include the following:

- *Clean gasoline.* In the United States, the amended 1990 Federal Clean Air Act imposed a tax on the U.S. refining industry that may ultimately cost $30–$60 billion. Western Europeans are going through the same revolution. In fact, they are trying to accomplish in 10 years what has taken the U.S. industry almost 30 years to achieve: moving from a leaded to an unleaded gasoline standard, mandating oxygenated fuels, and requiring the use of extremely low-sulphur diesel fuel for on-road vehicles.
- *Fuel economy.* For many years, improving the fuel economy of vehicles has been a matter of great interest to Western governments. In his book *Earth in the Balance: Ecology and the Human Spirit*, Vice President Al Gore discusses the desirability of raising the corporate-average fleet efficiency standard from 27.5 miles a gallon today to 40 miles a gallon within about five years.[1] Such a goal would increase the pressure on the oil industry.
- *Zero-emission vehicles.* In California, current law requires a certain percentage of automobiles sold by the end of the decade to be powered by storage batteries or solar power.
- *Site remediation.* This process restores the ground under closed service stations and refineries to their previous virginal conditions. It requires an enormous amount of money. Despite the current period of reduced downstream profitability, many refineries have not closed because restoring the site is just too costly. Eventually, refineries that should close will probably be reconfigured as product terminals, fuel oil or lube oil refineries, or other facilities.
- *Drilling bans.* The temporary ban on drilling for exploration and production in California, Florida, and North Carolina is likely to become permanent under the Clinton administration.

The Future of the Petroleum Industry

The oil industry should experience slow growth overall, about 1.3–1.8 percent a year in unit terms, with little growth in North America and Europe. Virtually all of the growth will be in the developing countries of Asia and Latin America.

The next major phase of the industry will be a pronounced shift from oil to natural gas. Natural gas (methane) has been used for home heating in the United States for almost 40 years, and it will increasingly displace coal, fuel oil, and nuclear power for power generation. In other parts of the world, the shift from heavy hydrocarbons to natural gas for power generation is only now getting under way. Europe, second to the United States with respect to natural gas intensity, relies on natural gas for about 12 percent of its primary energy (versus 25 percent in the United States). Tremendous gas growth lies ahead in Europe, as well as in such places as Asia, South America, and Mexico.

The downstream business will experience relentless pressure on margins because of the combination of slow growth and the environmental movement. The franchise value of surviving firms, however, will be enhanced. No one would go into the refining or marketing businesses on the basis of today's environmental requirements and profits; so firms that have the financial resources to survive this difficult period should be in a position to set prices after the shakeout.

[1]Albert Gore, Jr., (Boston, Mass.: Houghton Mifflin, 1992).

Table. 3 Ranking of Top 10 Oil Companies by Production and Refinery Capacity, 1991

		Production		Refining Capacity	
Company	Sales ($billions)	Millions of Barrels per Day	Rank	Millions of Barrels per Day	Rank
Saudi Aramco	—	5.3	1	1.6	8
Royal Dutch/Shell	$99.9	1.9	6	4.1	2
Exxon Corp.	90.3	1.8	7	4.2	1
PEMEX	16.2	2.9	2	1.7	7
PdVSA	11.1	2.0	5	1.9	5
National Iranian Oil Co.	—	2.9	3	0.8	14
Mobil Corp.	56.2	0.8	18	2.1	4
Chevron Corp.	32.8	0.9	13	2.2	3
British Petroleum	47.6	1.4	8	1.8	6
Kuwait Petroleum Co.	12.8	1.4	9	0.9	13

Source: Petroleum Intelligence Weekly.

Table 3 highlights the disparities between the production and refining capacities of the top 10 oil companies. Based on the 1991 annual survey by *Petroleum Intelligence Weekly,* Saudi Arabia's national oil company, Saudi Aramco, ranks first in production of oil but eighth in refining capacity. The only large company in reasonable balance is Petroleos de Venezuela (PdVSA), which after acquiring refining assets in the United States now ranks fifth in both world oil production and world refining. One of the most important trends in the next 10 years will be the righting of this imbalance. Those companies that have oil production capacity, primarily the Middle Eastern producers, will join forces with the former Seven Sisters, the former distribution arms, to access refining capacity. Formal agreements between these interests will become popular. One indicator of this trend is the successful formation of Star Enterprises several years ago; in that transaction, Saudi Aramco purchased half of Texaco's refining and marketing capacity east of the Mississippi—to the great advantage of both parties.

The oil industry has seen enormous change in its 134-year history. Through it all, the industry has shown great growth, profitability, and most important, resilience. Far from being in danger of extinction, the private-sector oil industry is more vital than at any time in history. As governments compete for votes in newly democratic states (e.g., Russia), financial and natural resources must be allocated more wisely than ever before. The result will be an inevitably larger role for the private-sector oil firms in the world of the 21st century. Well-capitalized and well-managed firms will prosper in the environment, as will their shareholders.

Question and Answer Session

Bernard J. Picchi, CFA

Question: Will OPEC's control of oil prices increase or decrease during the next three to five years? Is OPEC's long-term strategy based on other countries exhausting their resources first?

Picchi: OPEC produces only 37 percent of the world's supply. In the future, its control of oil will probably increase, although its control of oil prices may not. Many resource owners learned the hard way that controlling and setting price without regard to its impact on exploration or consumption invites problems. The more progressive OPEC countries—Venezuela, Saudi Arabia, and Kuwait—understand this connection. For example, Saudi Arabia's strategy today is to keep oil prices low and stable to gain greater control of oil. It wants to increase its share of the world oil market, which can be done only through stability and predictability of pricing.

Question: According to press reports, the emergence of North Sea oil diminished the brotherly cooperation among OPEC members. Does North Sea oil present a threat to OPEC's existence?

Picchi: North Sea oil is one of the primary reasons OPEC found it could not arbitrarily set oil prices based solely on its own revenue needs. OPEC increased the price of oil from an average of $2.60 a barrel in 1973 to $17 a barrel today but failed to maintain volume. During that period, non-OPEC oil production increased from 26 million barrels a day to about 40 million barrels a day. The North Sea was one of the major factors in that growth.

North Sea oil may not be as great a threat to OPEC as it once was, however. The North Sea today produces little more than 4 million barrels a day. The U.K. producers have suffered production declines throughout most of the past seven years, although Norway has continued to increase production. Combined total production in the North Sea will probably remain at about 4 million barrels a day.

For the past 10 years, OPEC has forecasted the imminent decline of non-OPEC oil production. Such forecasts demonstrate OPEC's failure to understand the economic motivation of non-OPEC producers to find oil.

Question: How will U.S. environmental compliance costs increase gasoline prices?

Picchi: The federal excise tax on gasoline is now 12 cents a gallon. The move to reformulate gasoline and to produce low-sulphur diesel fuel will increase the retail price of gasoline and diesel by 3–5 cents a gallon annually for the next two or three years.

Consumers are not sensitive, however, to price changes of 5 cents a gallon at the gas pump. On November 1, 1992, new requirements for the blending of oxygenated fuels went into effect in the 39 U.S. cities with the worst pollution problems. The result was a pump tax of about 5 cents a gallon. This tax went into effect in major metropolitan areas with hardly a whimper.

Question: If the upstream is so much more profitable than the downstream, why do the large, well-respected companies have so much refining capacity? In which direction is the industry moving?

Picchi: Before 1973, the only way the Seven Sisters had to move the oil they were then producing from their concessions in Saudi Arabia, Kuwait, Venezuela, and Nigeria was through company-owned refineries and service stations. In the 1970s, OPEC countries of the Middle East, Latin America, and Africa nationalized the oil concessions of the seven oil companies, but they could not nationalize their refineries because those assets were in the United States, Canada, Europe, and Asia. That was how the Seven Sisters became disproportionately larger players in refining and marketing than in production: Their concessions were seized. As it turns out, ownership of those distribution facilities was the key to controlling the market.

Venezuela was the exception to this model. Exxon, Shell, and other oil companies had built refineries in Venezuela; thus, the Venezuelans were able to nationalize refineries as well as oil reserves. Venezuela was also the only oil producer within OPEC that did not suffer a decline in income when oil prices collapsed in 1986. The other 12 members of OPEC looked at Venezuela and asked why that country had escaped the oil price collapse so well. The answer, of course, is that Venezuela owns its refineries; its production capacity is perfectly balanced with its refining capacity. In the future, we will see many deals between reserve-rich but downstream-poor OPEC producers and private-sector oil firms in the West.

Question: When do you expect to see a reversal in the decline of both oil production and consumption in the Commonwealth of Independent States? What will be the factors contributing to this reversal?

Picchi: The Russians want Western technology, but they do not want Western ownership of their oil resources. Russians take the issue of ownership of oil as seriously as the Mexicans. They have adopted the same attitude of national resource patrimony that is prevalent among Third World countries.

Power relationships in Russia have been much more important than money. Russian currency is worthless. The top social levels of Russian society have had benefits that money cannot buy—access to health care, automobiles, drivers, private schools, passports, and travel on demand. The presence of Mobil and Exxon personnel in the former Soviet Union would threaten the order of this closed, hierarchical society, especially for those at the top who are in a position to hurt Western investments in the Russian oil patch. Unless Russian society is completely reformed, therefore, the system will not welcome the absorption of much Western capital, and Russian oil production will continue to fall.

Question: Which producing areas of the world are likely to benefit from a gradual shift to natural gas?

Picchi: The largest owners of natural gas reserves in the world today are Russia, the Middle East (especially Iran, Saudi Arabia and Qatar), and the countries of western and northern Africa. Given the relative political stability of certain owners of Persian Gulf natural gas reserves, and their proximity to Asian markets, they probably will be the biggest beneficiaries of the trend toward natural gas.

Question: What companies are in the best position to succeed in the future?

Picchi: The keys to success in this business are discoveries, management strength and vision, and financial capacity. The major international companies are well positioned in today's world: Exxon, Mobil, Texaco, Shell, Total Petroleum, and Atlantic Richfield should do well in this environment.

Question: What are the prospects of nonfossil-fuel energy sources significantly displacing oil and natural gas as energy sources? What new technologies offer promise in this area in the next 10 or 20 years?

Picchi: I can't see real threats in the next 5–10 years. In the long run, photovoltaic and hydrogen-based power systems could threaten fossil fuels—as could electronic technologies that would make certain forms of travel obsolete (for example, video conferencing replacing plant visits).

Question: One of the most important trends you highlighted is the move to balance production and refining capacity. Given the low returns in the downstream, what is the incentive for companies and countries to do this?

Picchi: If owning the market results in higher crude sales, the strategy could be a winner, even if the "market" (i.e., refining capacity) is not a winning one. In practice, refineries have discovered two reasons that returns actually rise when they form affiliations with national oil producers: (1) Refineries achieve greater economies when they run one crude refinery, and (2) working capital is less expensive to finance, because the national oil company holds the crude inventory, which costs it nothing, until the crude is refined.

Factors Affecting Oil Industry Dynamics

William L. Randol
Director of Stock Research
Salomon Brothers, Inc.

Although oil companies are influenced more by external factors such as refining margins, geological prospects, and environmental policy than by internal factors such as reserve replacement ratios, cash flow, and earnings mix, both sets of factors play key roles in company success.

Companies in the oil industry are influenced more by external factors than by internal factors. Within each company, however, both sets of considerations play a role. Companies often make management decisions regarding internal factors based on external conditions. In turn, the actions of each company in the aggregate help create the overall industry environment. This presentation covers the major external and internal factors affecting the oil industry.

Key External Factors

The key external factors affecting the industry are the determinants of oil prices, refining margins, world economic growth and demand for oil, geological prospects, government regulation and environmental policy, and capital costs.

Determinants of Oil Prices

Many factors—supply and demand considerations, the energy futures market, future price expectations—affect the price of oil. This discussion focuses on supply and demand.

The price of oil is largely determined by OPEC, and the price of North American natural gas is largely determined by supply and demand.

Since the 1970s, OPEC has been the strongest influence on oil prices because of its dominance of world supply. Total world petroleum reserves are roughly 1 trillion barrels, and as shown in **Figure 1**, the Middle East OPEC countries of Iran, Iraq, Kuwait, Saudi Arabia, and the United Arab Emirates (UAE) control about 660 billion barrels, two-thirds of those proven reserves. Other OPEC countries control an additional 11 percent of the total.

Although Saudi Arabia officially controls 257 billion barrels in reserves, informed sources put the total closer to 300 billion after the recent discovery of 30–40 billion barrels of light crude in the southeastern part of the country. Saudi Arabia's total would thus amount to 45 percent of the Middle East's 660 billion barrels. Iraq, Iran, Kuwait, and the UAE each have roughly 15 percent, about 100 billion barrels.

Oil prices are weak today because OPEC is producing too much oil. **Table 1** shows the world supply-and-demand balance. The world is consuming approximately 67 million barrels of oil a day. Non-OPEC countries produce oil at their maximum economic rate given considerations of balance of payments and security of supply. OPEC, the residual supplier to the world, produces enough to balance demand. In the fourth quarter of 1992, OPEC produced an estimated 25.5 million barrels of crude a day with no inventory draw. At this high production rate, spare capacity in the system is very low. Saudi Arabia, for example, has a maximum of only 500,000 barrels of spare capacity a day, enough to meet 0.7 percent of current demand.

A comparison of U.S. and OPEC trends in crude oil production can be made from **Figure 2**. U.S. production has declined from about 9.1 million barrels of crude oil a day in 1986 to about 7.2 million barrels in 1991, not including natural gas liquids. Most experts predict production will continue to decline by 200,000–400,000 barrels a day annually. Prudhoe Bay, the largest U.S. oil field, peaked in 1987 at 1.5 million barrels a day. Without Prudhoe Bay, the average production per U.S. well is only 10–12 barrels a day. In comparison, wells in the Persian Gulf produce 20,000–40,000 barrels a day, which illustrates how the maturity of the U.S. domestic industry has led to use of less productive wells.

Figure 1. Distribution of Global Proven Petroleum Reserves, 1991

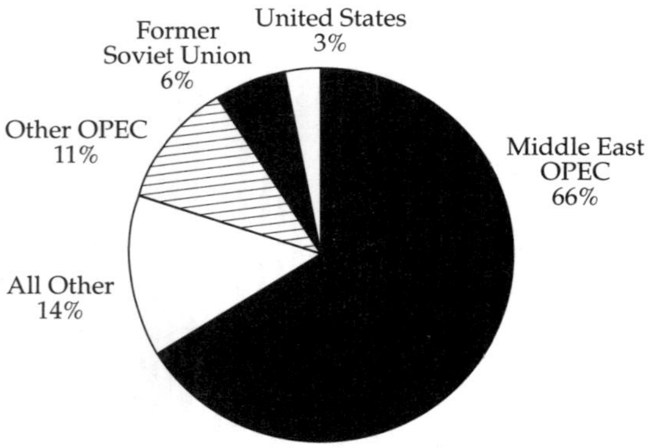

Source: Salomon Brothers, Inc.

Note: Total = 1,001,572,000 barrels.

Figure 2. OPEC and U.S. Crude Oil Production, 1980–91

Source: Salomon Brothers, Inc.

Between 1978 and 1985, OPEC lost half its market as production slipped from 31 million barrels to 15 million barrels a day. OPEC production has recovered somewhat since 1985; it grew at a rate of more than 1 million barrels a day annually to its current level of 25 million barrels a day.

Growth in demand for OPEC oil is a function of growth in world demand and growth in non-OPEC supply. **Figure 3** shows how much these growth factors increase or decrease demand for OPEC oil within a one-year horizon. Low growth in world demand and high growth in non-OPEC supply translate into weak oil prices, which is good for consumers. Conversely, high growth in demand and negative growth in non-OPEC supply mean a rise in prices, which is good for OPEC. The outlook for growth in OPEC's market through the mid-1990s is about 1 million barrels a day, annually, which corresponds to growth of about 1 percent annually in world oil demand after allowing for the decline in production in the former Soviet Union.

Refining Margins

Another important external factor is refining margins, which are related to levels of gasoline inventories. Simply put, raising gasoline prices is difficult when gasoline inventories are high. Companies increase inventories of gasoline in the early months of each year following the heating season and then deplete the inventories during the summer months, when automobile use increases. Inventories of 220 million barrels or lower at the end of the winter period usually signal a good summer driving season for the oil companies: Gasoline prices will be strong.

Figure 4 presents an example of the inventory-to-margin relationship by plotting U.S. gasoline inventories against Gulf Coast refining margins from 1989 to 1992. The two series tend to be mirror images of each other: Margin peaks usually correspond to lows in the inventory cycle and vice versa. If the oil companies forecast a good driving season, they gear up to make a lot of gasoline. If all the companies do this, however, overproduction causes a depression in gasoline prices.

Economic Growth and Demand for Oil

The relationship between oil use and economic activity has changed significantly during the past few decades. In the 1960s, a unit of economic growth generally corresponded to one or more units of additional energy to power this growth, which translated into a unit of oil consumption. Two major price disruptions in the 1970s—the Arab embargo in late 1973 and the Iranian revolution in late 1978—led to a change in this relationship. Since the mid-1980s, energy and oil consumption have stabilized at an empirical ratio of 1.00 to 0.50–0.75; that is, an addi-

Table 1. Short-Term World Petroleum Supply and Demand, 1991–2Q'93[a]
(millions of barrels per day)

Region	1991	1992 1Q	1992 2Q	1992 3Q	1992 4Q	1992 Year	1991–92 Change	1993 1Q	1993 2Q
Demand									
North America	18.6	18.7	18.5	18.9	19.2	18.8	1.10%	19.2	18.7
Europe	13.4	14.0	12.9	13.2	14.0	13.5	0.70	14.0	13.0
Pacific	6.1	6.7	5.7	5.7	6.4	6.1	0.00	6.8	5.8
Total OECD[b]	38.1	39.4	37.1	37.8	39.6	38.5	1.00	40.0	37.5
China, Asia, and former USSR	16.6	17.1	16.1	15.4	16.1	16.1		16.5	15.8
Eastern Europe	1.2	1.2	1.1	1.0	1.0	1.1		1.1	1.1
Latin America	5.3	5.1	5.4	5.6	5.6	5.4		5.3	5.6
Middle East and Africa	5.6	5.9	5.8	5.9	6.0	5.9		6.0	6.1
Total non-OECD	28.6	29.2	28.3	27.9	28.6	28.5		28.9	28.6
Total demand	66.7	68.6	65.4	65.7	68.2	67.0	0.60	68.9	66.1
Supply									
Total OECD	16.3	16.8	16.2	16.3	16.7	16.6		16.7	16.3
Former USSR	10.4	9.5	9.2	8.8	8.7	9.0		8.5	8.3
China	2.8	2.8	2.8	2.8	2.8	2.8		2.8	2.8
Other non-OECD	10.6	10.7	10.7	10.8	10.9	10.8		10.9	10.9
Total non-OECD	41.4	41.2	40.4	40.2	40.6	40.6		40.3	39.7
OPEC crude production	23.3	24.1	23.6	24.4	25.5[a]	24.4[a]		26.5[a]	
OPEC natural gas liquid production	2.1	2.1	2.1	2.1	2.1[a]	2.1[a]		2.1[a]	
Total OPEC	25.4	26.2	25.7	26.6	27.6[a]	26.5[a]		28.6[a]	
Total supply	66.8	67.3	66.1	66.8	68.2[a]	67.1[a]		68.9[a]	
Difference between total demand and total supply	0.1	–1.3	0.7	1.1	0.0[a]	–0.1[a]		0.0[a]	

Source: International Energy Agency.

[a]Estimate.

[b]Organization for Economic Cooperation and Development.

tional unit of economic growth now translates into between 0.50 and 0.75 of a unit of growth in oil demand. OPEC, being the residual supplier, lost half its market between 1979 and 1985. After that loss, some OPEC members (including Saudi Arabia and Iran), fearing a permanent reduction in demand for their product, decided that prices could never again be allowed to increase as they had in the 1970s.

Another demand factor contributing to today's low oil prices is the collapse in natural gas prices to about $1 per thousand cubic feet (mcf) in February 1992. Prices recovered in October to $2.74 and were about $2.40 in early 1993. In countries outside North America, natural gas prices sell at heating-value parity with oil. In the United States and Canada, natural gas prices are determined by supply and demand factors that do not always bear a relationship to the price of oil.

Although the future is uncertain, world consumption of oil should grow at about 1 percent a year—an aggregate of high growth in the Pacific Rim and Mexico, somewhat higher growth in the less developed countries and the OPEC nations, and negative or no growth in the United States, Canada, and other industrial countries. The former Soviet republics present an area of uncertainty in future production and consumption. Oil production in this region is plummeting. It peaked at about 12.2 million barrels a day in 1988 and has been falling more than 1 million barrels a day annually since then. Production is currently less than 9 million barrels a day. Consumption is also falling, however. Russia has been extremely inefficient in its use of energy. Under communism, people were more concerned with preserving their heating allocation than with conserving energy. Houses and buildings have neither insulation nor thermostats. To manage interior temperatures during the winter, people open windows. Assisting the Russian infrastructure to consume less energy will be a great growth industry in the future.

Figure 3. Growth in Demand for OPEC Oil
(millions of barrels per day)

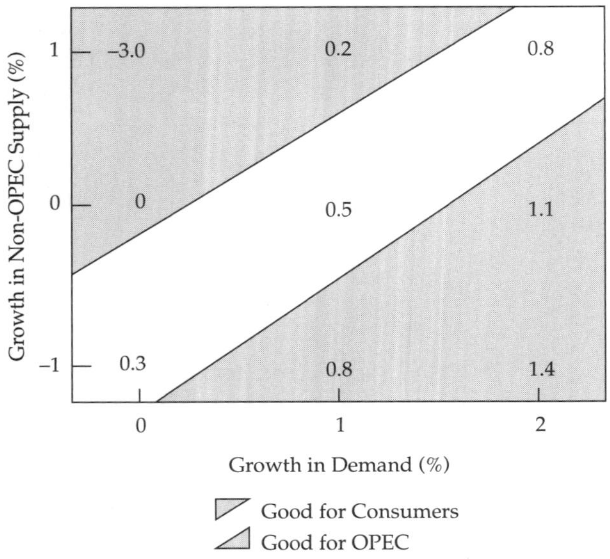

Source: Salomon Brothers, Inc.

Geological Prospects

Companies allocate capital for drilling according to what they judge are the best geological prospects. The United States is one of the most heavily drilled countries in the world. Companies continue to explore in the Gulf of Mexico, the site of some sizable gas and oil discoveries, but the more attractive drilling areas off California and in the Arctic National Wildlife Refuge are off-limits for environmental reasons. The result is a transfer of exploratory capital from the United States overseas, where the geological prospects and fiscal regimes are favorable and returns on investment are higher.

Government Regulation and Environmental Policy

The U.S. government has a market-oriented approach to the oil business. It is not proactive in encouraging companies to explore for oil and gas reserves. As Bernard Picchi mentions, the U.K. government softened the tax and royalty rates and created new exploration incentives when the price of oil fell in 1986.[1] The U.S. government has never done that. In fact, it has removed some incentives, which has encouraged exploration dollars to flee the country.

The U.S. government is requiring oil companies to increase their production of reformulated gasoline, and they are not happy about it. Although reformulated fuel may be good for the environment, it is not necessarily good for the oil industry. First, the use of methyl tertiary butyl ether (MTBE) and other oxygenates will increase the gasoline pool by about 7 percent. This increase in product will reduce capacity utilization rates, which may lower gasoline prices and create a squeeze on profit margins. Second, the companies view the capital investments required to produce reformulated fuel as zero-return investments. When the government's reformulated fuel program is in place by 1995, the per-unit additional cost for a gallon of gasoline will be between 15 and 20 cents a gallon. As the cost of the reformulated fuel works its way into the price structure, demand will decline and the industry will have difficulty recovering its substantial new investment.

[1] See Mr. Picchi's presentation, pp. 3–10.

Figure 4. Total U.S. Gasoline Inventories and Gulf Coast Refining Margins, 1989 to October 16, 1992

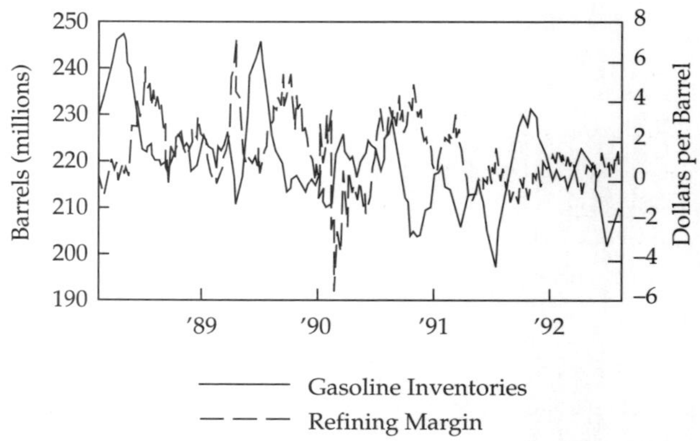

Sources: American Petroleum Institute, *Weekly Statistical Bulletin; Petroleum Intelligence Weekly; Reuters Benchmark.*

Capital Costs

The cost of capital is another important external factor. In general, returns on investment have been below the cost of capital. The trend among the Seven Sister companies (as **Table 2** shows) is to invest outside the United States, where the returns on investment have typically been greater than in the United States.

Key Internal Factors

The key internal industry dynamics affecting oil and gas companies are reserve replacement ratios, finding costs, production profiles, earnings mix, return on investment, cash flow, debt ratios, and allocation of capital.

Reserve Replacement Ratios

The general shift of U.S. oil company capital

Table 2. Non-U.S. Expenditures by the Seven Sisters as a Percent of Total Expenditures, 1985 and 1991

	Capital Expenditures		Exploratory Expenditures	
Company	1985	1991	1985	1991
British Petroleum	41%	77%	35%	66%
Chevron Corp.	28	47	42	65
Exxon Corp.	48	65	34	64
Mobil Corp.	35	56	50	75
Royal Dutch/Shell	40	77	59	77
Texaco	28	46	34	67

Sources: 1991 annual reports; Salomon Brothers, Inc.

away from domestic investment and toward development of overseas holdings is evident in the differences in reserve replacement ratios over time. **Figure 5** shows U.S. and non-U.S. (international) reserve replacement ratios for the Seven Sisters during the past five years. Reserve replacement ratios in the United States have generally declined, but internationally, except for British Petroleum, the companies have replaced more than they produced. In 1991, for example, Royal Dutch/Shell failed to replace reserves in the United States because of the poor economics of doing so, but its reserve replacement ratio outside the United States was greater than 100 percent.

Finding Costs

After reserve replacement ratios, finding costs are the most critical internal factor for the oil companies. The more effectively a company finds new oil fields, the lower its finding costs. Although finding costs vary from company to company, they are generally between $4 and $9 a barrel.

Production Profile

The key to a company's economic health is how much oil and gas it can produce. Equity oil and gas production offers attractive margins because profits are a function of the price of oil. In nonequity production, producers are paid a fixed amount on every barrel they produce regardless of the price of oil. Equity production is the lifeblood of the earnings and cash flow of oil companies.

U.S. and international oil and gas production figures for the Seven Sisters from 1987 through 1991 (as a ratio of their 1987 levels of production) are shown in **Figure 6**. Gas production in the two graphs of the figure has been converted to an oil equivalent on the basis of its heating value, 6 mcf a barrel. Most companies use this method of conversion, even though it does not reflect pricing disparities.

Only Chevron and Mobil produced more U.S. oil and gas in 1991 than they did in 1987. Royal Dutch/Shell, with the lowest reserve replacement ratio in the United States, also has the lowest relative production level. In 1991, Shell produced about 80 percent of what it had produced in 1987. Texaco performed the worst outside the United States; its 1991 production was only about 68 percent of what it was in 1987. Exxon's non-U.S. production also fell in 1991 to average about 75 percent of 1987's level.

Earnings Mix

The oil companies' sources of earnings have changed dramatically during the past six years away from the United States toward international markets, which have superior returns. **Table 3** compares the non-U.S. component of the Seven Sisters' earnings as a percentage of their total earnings in 1985 and 1991. British Petroleum has gone from 38 percent non-U.S.

Figure 5. U.S. and International Reserve Replacement Ratios for the Seven Sisters, 1987–91

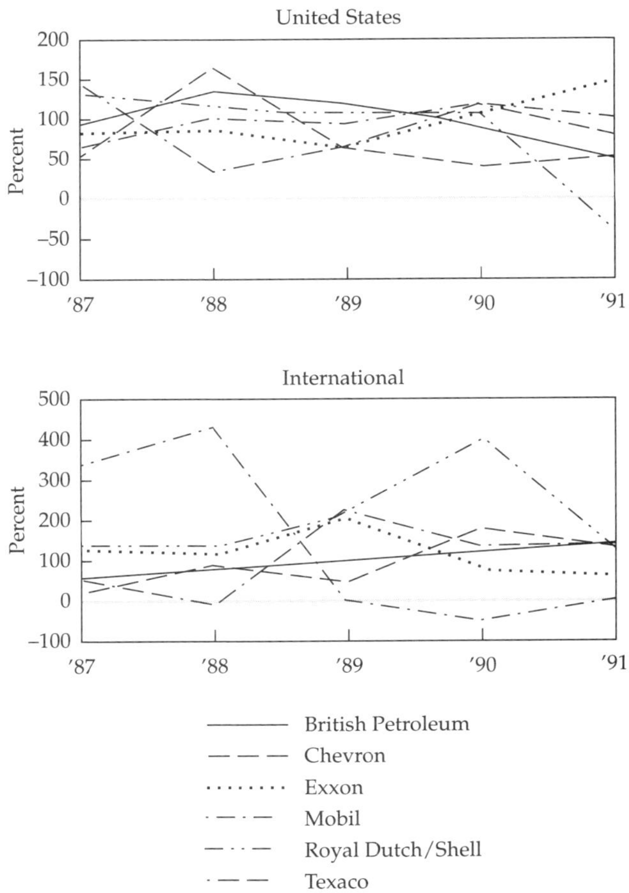

Sources: 1991 annual reports; Salomon Brothers, Inc.

Figure 6. Ratio of the Seven Sisters' U.S. and International Net Oil and Natural Gas Production for 1987–91 to 1987 Production

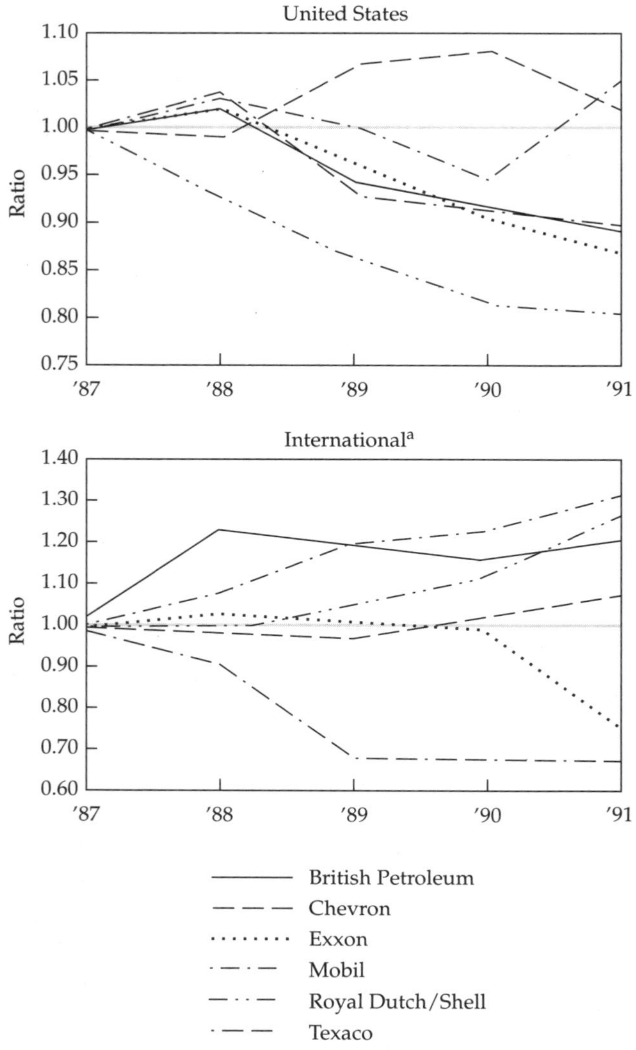

Sources: 1991 annual reports; Salomon Brothers, Inc.

[a]Figures for Texaco prior to 1989 have not been adjusted to reflect the sale of Texaco Canada to Imperial Oil in 1988.

earnings to 59 percent, Chevron's non-U.S. earnings mushroomed from 35 to 83 percent, and Mobil's increased from 57 to 86 percent. Royal Dutch/Shell, which had large write-offs in the United States, increased its international earnings from 66 to 100 percent.

Return on Investment

The price of oil has been on a roller coaster since 1970. Not many years ago, the price of Saudi Arabian light crude was $34 a barrel. When oil prices collapsed in 1986, the price of Middle East crude fell to $7 a barrel. Although the price has since recovered to $18 a barrel, oil companies made significant capital investments when the price of crude was $34 a barrel, and returns have suffered as a result.

Table 4 shows U.S. returns on investment for six oil companies according to source: upstream production, downstream production, or chemicals. Except for Arco Chemical, which is the most efficient refiner/marketer in the country, the returns in the oil segments of the business have been extremely poor. Exxon was the only company other than Arco for which the six-year average return on investment in the upstream segment was better than 5 percent. Exxon achieved that level because of its 20 percent position in Prudhoe Bay production; Prudhoe Bay is the most profitable place to operate in the United States.

Returns in the downstream sector were also poor, although they were better than the upstream returns. Chemical earnings were high in the late 1980s, but the combination of overcapacity and recession pushed them down to almost zero by the end of 1992.

Cash Flow

Cash flow is the single most important measure when valuing oil companies, because cash flow is more stable than earnings. Cash flow in the oil industry is defined as earnings after taxes plus depreciation, dry hole expenses, and deferred taxes. Cash flow multiples can be used in combination with price-to-earnings multiples for valuation. Typically, oil companies sell at anywhere from 4.5 to 7 times cash flow. Exxon, for example, was recently downgraded on the basis of cash flow valuation; Exxon stock is selling at 7.1 times cash flow while most other companies in the group are selling at 5 times cash flow.

Debt Ratios

Debt ratios are important in valuation, and they

Table 3. Non-U.S. Earnings by the Seven Sisters as a Percent of Total Operating Earnings, 1985 and 1991

Company	1985	1991
British Petroleum	38%	59%
Chevron Corp.	35	83
Exxon Corp.	64	76
Mobil Corp.	57	86
Royal Dutch/Shell	66	100
Texaco	47	58

Sources: 1991 annual reports; Salomon Brothers, Inc.

Table 4. U.S. Returns on Investment—Six Oil Companies, 1986–91

Company	1986	1987	1988	1989	1990	1991	Six-Year Average
U.S. upstream							
Arco	2.8%	11.3%	9.5%	14.3%	19.0%	11.5%	11.4%
Chevron Corp.	0.5	3.0	(0.2)	3.5	9.0	3.8	3.3
Exxon Corp.	4.0	7.7	4.8	6.8	7.7	4.0	5.8
Mobil Corp.	1.5	4.1	1.3	2.2	4.3	2.6	2.2
Shell	3.1	5.2	2.4	4.1	5.4	1.8	3.7
Texaco	(0.7)	2.2	1.7	4.5	10.3	7.9	4.3
U.S. downstream							
Arco	30.2	11.4	25.0	18.0	24.3	12.9	20.3
Chevron Corp.	5.1	1.9	14.0	5.4	9.2	3.2	6.4
Exxon Corp.	13.5	0.9	17.9	9.5	1.9	12.6	9.4
Mobil Corp.	8.4	2.9	13.5	8.2	5.1	4.5	6.8
Shell	8.6	6.1	13.9	9.4	2.2	(3.9)	6.0
Texaco	3.1	(1.4)	25.6	3.9	11.7	8.5	8.6
U.S. chemicals							
Arco	9.1	46.2	38.8	28.7	17.9	9.8	25.1
Chevron Corp.	15.7	31.3	59.5	38.0	15.4	8.7	28.1
Exxon Corp.	13.1	20.2	39.1	32.3	15.3	13.8	22.3
Mobil Corp.	6.0	17.4	33.0	27.8	11.0	2.1	16.2
Shell	18.4	22.7	29.7	32.7	16.4	7.4	21.2
Texaco	3.7	3.9	31.2	35.0	3.0	(0.2)	12.8

Sources: 1991 annual reports; Salomon Brothers, Inc.

Note: Arco's upstream income and investment numbers reflect Alaska only.

vary greatly from company to company. In 1990, British Petroleum's debt was 40 percent of total capital, which is high for this industry. Today, British Petroleum is suffering the consequences of that debt level. The economic downturn caused industry fundamentals to deteriorate, and when British Petroleum's high debt service affected its earnings and cash flow, it was forced in August 1992 to cut its dividend. The company also drastically reduced capital and exploratory spending and put $1.5 billion to $2 billion worth of assets on the auction block, with plans to do so every year for the next few years.

Although Arco, Phillips Petroleum, and Unocal Corporation have relatively high debt ratios, Royal Dutch/Shell is at the other end of the spectrum with a ratio of about 10 percent debt to total capital. If Royal Dutch/Shell were to increase debt to 65 percent of total capital, it could theoretically generate $65 billion in cash. With that much cash, it could buy Exxon.

Allocation of Capital

The trend in allocation of capital, whether in the upstream, downstream, or chemicals segments, is away from the United States. Table 2 showed the extent of this capital flight for the Seven Sisters. In 1985, the major oil companies made less than half of their capital expenditures overseas; in 1991, the percentage of capital allocated overseas was as high as 77 percent. This growth in international capital spending is reflected in increases for all the companies in percentage of total net property, plant, and equipment located elsewhere. Nearly three-quarters of Royal Dutch/Shell net PP&E is now located outside the United States.

Conclusion

This industry has many different facets that combine to create the overall industry environment. Mobil provides a good example of how internal and external factors affecting the oil industry weave together. Mobil cut its capital spending three times in 1992 because of a neutral or pessimistic outlook on crude oil prices, refining margins, and the chemical industry. Its actions illustrate how external factors influence internal factors and, most importantly, management decisions.

Question and Answer Session

William L. Randol

Question: How would a strong U.S. dollar affect the oil companies' profitability?

Randol: Crude is denominated in U.S. dollars. With a stronger dollar, Exxon's German affiliate would pay more deutsche marks to buy a barrel of crude to refine. This squeezes the affiliate's downstream margins unless it can raise the price of gasoline and other petroleum products to reflect the increased cost. For upstream operations outside the United States, a strong dollar is beneficial. For example, one year ago, the ratio of the U.S. dollar to the British pound was 2 to 1. It is now, reflecting a strengthening dollar, $1.53 to the pound. Because crude is denominated in dollars and the dollar has increased in value from last year, producers in the United Kingdom are now paid more revenue in pounds. That will benefit upstream earnings and operations outside the United States. A weak dollar helps downstream operations and hurts upstream operations outside the United States.

One other element is dividend considerations. On August 6, 1992, British Petroleum cut its dividend in half, to 2.1 pence per ordinary share—that is, 55 cents per American Depositary Receipt, or $2.20 annualized. With the change in the exchange rate, the dividend is now worth approximately $1.50. So the change in the dollar can affect dividends of foreign-based companies.

Question: When do you expect to see a reversal in the decline of oil production and consumption in the Commonwealth of Independent States? What will be the factors contributing to this reversal?

Randol: We do not have complete information regarding the former Soviet Union, so it is difficult to say when production and consumption will increase. The decline in consumption may bottom out sooner than the decline in production. The price of oil in the former Soviet Union has been moving up from an extremely low level and is now about $3 a barrel. I am not optimistic that Russian production will stabilize before 1995. Longer term, there is the prospect of boosting production.

The former Soviet republics have two major problems to overcome. One is payment. Because of perceived political risk, companies do not want to risk their capital in the republics, but these countries do not have any capital to contribute to the development of the oil. Companies that do contribute capital are very concerned about repatriation of profits. Hard currency is preferable but not possible, so payment is made in oil. Companies have to analyze their investments in terms of oil they hope to export on the world market for a reasonable price.

The other problem is education. The former Soviet republics are not familiar with many Western business concepts—return on investment, risk and reward, payment for services rendered, capital invested, and so forth.

Chevron's Tengiz Project in Kazakhstan, the most visible project in the former Soviet republics, is fraught with logistical and financial problems. The republic did not have any accounting books when Chevron bought into this venture and so had no record of the cost of building the plant. In addition, the desulphurization plant currently operating was built with bartered goods, and the crude at the plant is very high in sulfur content.

How Chevron will get this crude to the world market—or even whether the project will go forward—remains unclear. To make the project worthwhile, Chevron must export the oil to the world. The oil is on the northeast coast near the Caspian Sea, but the Caspian Sea is an unnavigable marsh, and tankers cannot go in there. Transporting the oil to the Black Sea would require building a pipeline through another republic. The pipeline could go north through the Russian Republic, but the Russians will demand compensation. To the south, there is fighting between the Azerbaijanis and other ethnic groups. Chevron has considered piping it south to Iran.

As of January 1, 1992, Chevron had invested five years in the project. It has the right to walk away if a second desulphurization plant is not up and running by January 1, 1993, but the plant will undoubtedly be operating by that time.

The former Soviet republics have an enormous resource base, but it will take a lot of capital, management, and technical expertise to develop it.

Question: Do the oil companies worry about nationalization of assets as they transfer capital overseas?

Randol: Nationalization is a concern. Oil companies address the concern through the political risk premium they assign to their cost of capital when they evaluate overseas projects as part of their capital allocation process. During the past decade, a tremendous change has taken place in the producing nations, including the hostile countries or previously hostile countries (such as Iran). They now recognize the benefits of price stability and are seeking Western partners. Even Iraq is talking about welcoming the major oil companies back after the trade sanctions are lifted. Yemen has a large amount of political risk, yet Chevron and Exxon are very profitable there.

Question: When will Iraq reenter the oil market significantly? Will other OPEC members such as Saudi Arabia accommodate Iraq's reentry to prevent an oil price collapse?

Randol: The Iraqis and the United Nations have met on three separate occasions to discuss resumption of Iraqi oil exports. What the President will recommend to the UN Security Council is not clear, but $1.6 billion in exports during a six-month period, or 500,000 barrels a day, has been discussed. OPEC could accommodate this volume. The Iraqis continue to regard the UN resolutions as an insult to their sovereignty. According to the UN resolutions, the revenues are to be allocated and disbursed by UN authorities for humanitarian purposes; Saddam Hussein would not receive any of the revenues.

At its February 1992 meeting, OPEC passed a resolution to convene an emergency meeting when and if Iraq and the United Nations reach an agreement to resume exports. The idea was that they would all cut production by 2 percent to accommodate the expected increase of 500,000 barrels a day.

Accommodation will not be a problem if the final UN resolution approximates the volumes discussed at the previous three meetings with Iraq. In late summer and early fall 1992, the Iraqis were talking about $4 billion worth of exports over six months, which would be 2.5 times the volume discussed previously. The new volume would cause a major absorption problem in the world supply-and-demand balance, particularly for OPEC. To maintain stable prices in such an environment, someone would have to cut production, because Iraq's reentry into the market would exacerbate the seasonal problem.

Question: Please discuss the differences between finding costs, development costs, and production costs.

Randol: Finding costs are the exploratory expenses incurred in discovering new oil and gas reserves but not in developing them. Development costs are the costs to develop reserves that have already been discovered and booked. Production costs are the actual out-of-pocket costs involved in physically producing a barrel of oil.

Question: Why has Texaco's production fallen so dramatically since 1987? Does this fall reflect the formation of Star Enterprises?

Randol: The drop in Texaco's worldwide oil production reflects the sale of Texaco Canada to Imperial Oil, a unit of Exxon. This drop primarily reflects the loss of Texaco's Canadian production. It has nothing to do with the formation of Star Enterprises.

Question: What are the political implications of the gradual shift of U.S. capital to foreign operations? What are the implications for U.S. oil companies? U.S. consumers? Foreign oil-producing countries?

Randol: U.S. oil companies are merely responding to the economic facts of life in this country and abroad by tilting their capital and exploratory spending overseas. By avoiding the low returns on investment that have characterized the U.S. upstream industry, the companies are merely doing what is in the best interests of shareholders. U.S. consumers will benefit to the extent that foreign oil supplies are cheaper to develop than domestic reserves. Foreign oil-producing countries will obviously benefit from this trend.

Question: The increase in non-U.S. capital expenditures seems to be significantly higher (in percentage terms) than the shift in PP&E. What major components account for the remaining big swing?

Randol: Capital expenditures are the leading edge in a company's total investment. Net PP&E is by definition a measure that represents the accumulated investment over time in a given business sector, net of accumulated depreciation. Capital expenditures can vary significantly from year to year, net PP&E is a massive figure that changes only very slowly.

Question: How much of the decline in U.S. production is attributable to political, cost, and other restrictions on exploration rather than purely geological considerations?

Randol: At current price levels, the decline in U.S. production simply reflects the economic realities of exploring for and producing oil in the United States. The United States is one of the most heavily drilled oil provinces in the world, and the domestic petroleum industry is mature by any standard. It doesn't help that some of the most attractive geological prospects have been withheld from the oil industry for environmental reasons.

Question: To what extent are the small independents moving capital overseas?

Randol: The economic forces driving investments overseas and away from the United States are the same for small independent companies as they are for the majors.

Question: Why do some companies have higher-than-average finding and development costs and lower-than-average production costs, or vice versa?

Randol: Companies with lower-than-average production costs could well be participants in large, mature oil fields, such as the Prudhoe Bay oil field, where economies of scale keep unit operating costs low. Whether a company has high or low finding costs is a function of how efficiently that company is spending capital to explore for oil and gas—that is, how good an oil finder it is.

Interpreting the Oil Industry Numbers

Michael L. Mayer
Director
Wertheim Schroder & Company, Inc.

The critical factors analysts should probe in determining whether to recommend purchase of oil stocks are the expected prices of oil and natural gas, worldwide downstream margins, and chemical margins. From these forecasts, a model will yield expected dividends, earnings, and cash flows, which will allow analysts to examine the effect on earnings of changes in the critical factors.

Oil industry analysts spend a lot of time on the numbers in the hope that, if they correctly forecast the earnings, cash flow, dividend growth, and asset value of stocks in the oil universe, the results will have some positive correlation with the stock's market performance and they will make money at the end of the day. This presentation focuses on the key factors in evaluating and comparing companies.

The heart of understanding and forecasting the earnings of oil companies lies in analyzing the upstream and downstream segments. **Table 1** shows the distribution of estimated 1993 operating earnings among the major integrated oil companies. Based on 1993 estimates of $20 a barrel for West Texas Intermediate (WTI) oil and an increase in the average price of natural gas from $1.65 to $1.80 a thousand cubic feet (mcf), the industry should experience a small recovery in the downstream segment and a small improvement in chemical profits.

As a group, these companies should generate about 77 percent of earnings, excluding special items, from the upstream segment. Approximately 43 percent will come from the downstream segment—27 percent in the United States and 16 percent outside the United States. The chemical sector contributes about 12 percent. Corporate overhead, interest expense, real estate operations, minerals, coal, and so forth account for a 32 percent charge against earnings.

Upstream Analysis

Forecasting revenues for the upstream segment (exploration and production) is simply a matter of correctly guessing oil and natural gas prices. Upstream revenues are easy to understand. The price of a barrel of oil multiplied by the volume produced will yield gross upstream revenues. The companies simply produce as much as they can.

Although costs receive less emphasis than revenues in industry analysis, they are the key to differentiating among stocks and among company managements. Oil industry executives cannot do much to affect the price of natural gas or oil—they are price takers—but they can control their costs during the intermediate term. Differentiating among companies by cost structure can thus lead to interesting conclusions.

Key variables in analyzing this industry include the following:

- *Oil and gas prices.* Generally, the higher the prices, the better for the industry. In some parts of the world, however, prices do not matter because companies operate on fixed margins. For example, companies producing oil in Nigeria generally receive about $2.25 for each barrel they produce regardless of whether the price of oil is $15 or $30 a barrel.
- *Oil and gas operating costs.* The trend toward higher costs is constant and inexorable. Analysts expect the price increases built into long-term models to be partially offset by higher operating costs.

Production costs are an important factor. Producing oil is much cheaper in Texas, which is also closer to the distribution and refining system, than it is in harsher environments such as the North Sea or the North Slope of Alaska.

- *Exploration costs.* A large integrated oil company such as Exxon will spend about $7 billion a year on exploration. If the exploration is successful, the company capitalizes the costs and charges them

Table 1. Major Oil Companies' Distribution of Operating Earnings, 1993 Estimate
(West Texas Intermediate oil at $20 per barrel; U.S. natural gas at $1.80 per mcf)

	Upstream		Downstream			
Company	U.S.	Foreign	U.S.	Foreign	Chemical	Other
Amoco Corp.	45%	17%	33%	0%	18%	–13%
Chevron Corp.	38	38	19	12	7	–14
Exxon Corp.	19	53	7	22	11	–12
Mobil Corp.	25	56	14	21	8	–24
Texaco	47	35	27	34	3	–46
Royal Dutch	6	56	4	41	5	–12
Shell	6	56	4	41	5	–12
Arco Chemical	55	6	35	0	31	–27
Phillips Petroleum	53	31	30	0	29	–43
Unocal Corp.	61	52	38	0	11	–62
USX-Marathon Group	67	26	90	0	0	–83
Group average	38%	39%	27%	16%	12%	–32%

Source: Wertheim Schroder & Co., Inc., estimates.

against earnings on a unit-of-production basis. If the exploration is unsuccessful, the company charges the cost immediately against earnings. Unsuccessful wells can thus distort quarterly earnings comparisons as the companies write off the associated drilling costs.

Finding and development costs. Analysts and investors sometimes misunderstand commonly cited statistics such as finding and development costs and reserve replacement ratios. They believe a company that adds oil and gas reserves to its books at a lower cost than its competitors is doing a better job. The belief is only true, however, if all barrels have equal profitability. They do not. Quality differences among crude oils contribute to price disparities. Obviously, adding low-quality oil reserves is not as profitable as adding reserves in areas where the quality and price of oil is high.

Taxes. In the United States, the marginal tax rate is the statutory tax rate. If the price of oil increases $1 a barrel, most companies will add about 60 cents a barrel to the bottom line; about 40 cents will go to the government in taxes. Some states apply additional taxes. Outside the United States, marginal tax rates average about 70 percent; in some areas, they are as high as 90 percent.

In the belief that more units times price will lead to higher earnings, some analysts look for companies with rising trends in production. More units do not necessarily lead to higher earnings, however. The profitability and cash flow characteristics of the new oil and natural gas must be comparable to or better than the normal decline in the existing production base.

How much cash flow companies can produce from operations is what really matters. **Table 2** presents data on cash flow for major oil companies from 1987 to 1991. Cash flow per barrel of oil equivalent (BOE) is after-tax earnings plus depreciation expense, adjusted for the after-tax effect of exploration spending, and divided by the BOE production.

Table 2 illustrates some errors analysts make by assuming earnings from production will be higher as oil and natural gas prices increase. In 1987, when the average price of WTI oil was $18.25 a barrel, European natural gas (a proxy for all natural gas outside the United States) sold for $2.45 an mcf, and U.S. natural gas sold for $1.67 an mcf; the average company in the industry generated $6.59 a BOE in cash flow. Analysts who were using an earnings model that assumed oil prices would increase 10 percent between 1987 and 1991 and that production levels would remain constant would have predicted increased cash flow and earnings in 1991. Actually, the average cash flow was only $6.12 in 1991, a decline of 7 percent, despite a 10 percent rise in oil prices on a worldwide basis and a rise of natural gas prices outside the United States of 31 percent. (The only laggard was U.S. natural gas prices, which declined 10 percent.) For these companies, however, the average cash flow per barrel declined 47 cents, and $1.50 a barrel in earnings and cash flow disappeared into the system.

Production Costs

Earnings failed to rise as the model would have predicted because of rising production costs. The average after-tax net income for the industry in 1991 was only $1.80 a barrel, down 18 cents from 1987. An analyst who in 1987 correctly predicted 1991 prices

Table 2. Major Oil Companies' Worldwide Cash Flow, 1987–91
(dollars per BOE, except as noted)

Company	1987	1988	1989	1990	1991	1991 versus 1987 Dollars/BOE	1991 versus 1987 Percent
Amoco Corp.	$ 6.28	$ 5.60	$ 5.97	$ 6.80	$ 5.54	($0.74)	–12%
Atlantic Richfield	5.94	5.09	6.08	7.10	6.04	0.10	2
British Petroleum	6.63	5.41	5.36	6.35	7.12	0.49	7
Chevron Corp.	5.82	5.06	6.13	7.25	5.65	(0.17)	–3
Conoco	5.58	4.46	4.80	6.44	5.33	(0.25)	–4
Exxon Corp.	6.68	6.01	6.61	8.45	7.05	0.37	6
Kerr-McGee	8.63	8.03	7.42	8.46	7.23	(1.40)	–16
USX-Marathon Group	8.04	6.44	7.49	8.87	7.24	(0.80)	–10
Mobil Corp.	7.56	6.78	6.12	6.84	5.88	(1.68)	–22
Phillips Petroleum	6.07	5.43	5.83	6.44	5.06	(1.01)	–17
Royal Dutch/Shell	5.94	5.77	5.25	6.16	5.36	(0.58)	–10
Texaco	5.84	4.48	5.35	6.62	5.60	(0.24)	–4
Unocal Corp.	6.63	6.01	6.57	7.50	6.47	(0.16)	–2
Average	$ 6.59	$ 5.74	$ 6.08	$ 7.18	$ 6.12	($0.47)	–7%
Posted WTI (dollars per barrel)	$18.25	$15.40	$18.60	$23.30	$20.15	$1.90	10%
European natural gas (dollars per mcf)	2.45	2.35	2.25	2.90	3.20	0.75	31
U.S. natural gas (dollars per mcf)	1.67	1.68	1.71	1.72	1.50	(0.17)	–10

Source: Wertheim Schroder & Co., Inc.

would have thought the average earnings after tax would be $3.30 a barrel. The analyst would have been off by a factor of 50 percent, because production costs went up.

Production, or lifting, costs include all costs incurred to extract oil from the ground—labor, repairs, maintenance, fuel consumed, property taxes, cost of wells, and so forth. Production costs averaged $4.38 a barrel for the industry in 1987, stayed relatively flat through 1989, and then increased in 1990 and 1991. For the five-year period, costs rose $1 a barrel, or 5 percent a year.

The increase in costs occurred for two reasons. First, production from large, mature U.S. fields has been declining, and oil companies are spending a lot of money trying to offset those declines. As the companies invest more capital in existing fields, average production costs increase. The second cause of rising costs, especially in the North Sea, is a worldwide increase in taxes and in environmental and safety regulations.

Production costs differ among the companies. Phillips Petroleum, for example, produces a lot of its oil in the North Sea, a hostile environment in which increasing operating costs, safety measures, and environmental regulations are squeezing earnings. The institution of a severe carbon tax by the Norwegian government in 1991 also raised Phillips's production costs and hurt its earnings. Phillips's production costs in 1989 were $3.81 a barrel; they jumped to $4.22 a barrel in 1990 and to $5.23 a barrel in 1991. During the two-year period from 1989 to 1991, production costs rose $1.50 a barrel, about equal to the total after-tax earnings on barrels sold.

The big Alaskan producers—Exxon, British Petroleum, and Arco Chemical—showed increases in production costs of about 8 percent a year, nearly twice the industry average. They made heavy expenditures to maintain production at Prudhoe Bay, the largest U.S. oil-producing field. In 1990, they installed a gas-handling facility at Prudhoe Bay to reinject the gas produced there to enhance oil recovery. Because about 42 percent of Arco's oil production comes from Prudhoe Bay, an increase in costs there has a disproportionate negative effect on Arco's earnings per barrel.

Depreciation

The companies' depreciation, depletion, and amortization (DD&A) expenses averaged $3.58 a barrel in 1991, as shown in **Table 3**. When companies are successful in bringing in new wells, they add the cost of exploration and development to the capitalized costs on their balance sheets and then depreciate those costs on a BOE basis as they produce oil. For example, a company might spend $100 million to explore and develop a field with all the necessary infrastructure. If the engineers estimate the field holds 33 million barrels, as the company produces oil throughout the year, it charges $100 million divided by 33 million barrels, or about $3 a barrel, against income as depreciation. This figure is added back to cash flow.

All companies say they are conservative in their accounting, but a comparison of depreciation rates and what remains on the books to be depreciated shows some large anomalies that affect future earnings. **Table 4** presents the major companies' capitalized costs, a proxy for future depreciation expenses. The average of capitalized costs on the books to be charged against earnings was $4.01 a barrel in 1991, more than the average depreciation rate shown in Table 3. To zero out the books according to the accounting rules, depreciation rates for the industry must increase about 40 cents a barrel, which squeezes earnings.

Forty cents a barrel may not sound like much, and if that were the figure for all the firms in the industry, analysts could probably forget about it, but the companies differ widely. For example, Phillips, based on this particular indicator, seems to be in excellent condition. In 1991, its year-end capitalized costs were $1.94 a barrel, but it charged $2.66 a barrel for DD&A against earnings. During the next several years, as the depreciation rate and the capitalized costs come into equilibrium, Phillips's pretax earnings could benefit by about 70 cents a barrel.

In contrast, British Petroleum (BP) had $6.05 a barrel of capitalized costs on its books at year-end 1991, and it charged only $3.83 a barrel for DD&A against earnings that year. If it correctly accounts for its book costs over time, the depreciation rate will increase $2.22 a barrel. If the price of oil increases $2.22 a barrel during the next two years, BP's earnings from existing developed reserves will not change because its depreciation charges must go up by an equal amount to zero out the books.

Figure 1 shows how much oil and natural gas prices must rise during the next five years to account for two likely factors: a 5 percent increase in annual production costs, and a depreciation rate equal to the capitalized costs left on each company's books. The figure assumes a five-year period to move from the current state of disequilibrium to a balanced state in which the books are in alignment. Unocal and Phillips appear to have significant competitive advantages by this measure. The price of oil and natural gas on a weighted basis will have to rise only about 70 cents for their earnings with existing production to remain unchanged. The industry average is $1.90,

Table 3. Major Oil Companies' Worldwide Depreciation, Depletion, and Amortization Expenses, 1987–91
(dollars per BOE)

Company	1987	1988	1989	1990	1991	1991 versus 1987	Annual Change, 1987–91
Amoco Corp.	$3.58	$3.48	$3.32	$3.41	$3.03	($0.55)	–4%
Atlantic Richfield	3.29	3.17	3.37	3.24	3.34	0.05	0
British Petroleum	2.94	3.35	2.91	3.15	3.83	0.89	7
Chevron Corp.	3.56	3.43	3.52	3.61	3.36	(0.20)	–1
Conoco	4.51	3.94	3.59	3.73	3.68	(0.83)	–5
Exxon Corp.	3.02	3.30	3.51	3.90	3.55	0.53	4
Kerr-McGee	6.39	6.26	5.59	5.29	5.25	(1.14)	–5
USX-Marathon Group	6.06	5.70	5.47	5.46	5.30	(0.76)	–3
Mobil Corp.	3.73	3.70	3.34	3.25	2.98	(0.75)	–5
Phillips Petroleum	3.64	3.12	2.86	2.71	2.66	(0.98)	–8
Royal Dutch/Shell	3.25	3.64	3.24	3.11	2.97	(0.28)	–2
Texaco	3.79	3.18	3.13	3.06	2.88	(0.91)	–7
Unocal Corp.	3.82	3.45	3.51	3.88	3.65	(0.17)	–1
Average	$3.97	$3.82	$3.64	$3.68	$3.58	($0.39)	–3%

Source: Wertheim Schroder & Co., Inc.

Table 4. Major Oil Companies' Worldwide Year-End Capitalized Costs, 1987–91
(dollars per developed BOE)

						1991	
Company	1987	1988	1989	1990	1991	U.S.	Non-U.S.
Amoco Corp.	$2.55	$2.95	$2.87	$2.80	$2.87	$2.37	$3.84
Atlantic Richfield	2.86	3.10	3.05	3.01	3.17	2.76	8.73
British Petroleum	3.81	4.81	4.72	5.64	6.05	3.36	9.35
Chevron Corp.	3.04	3.36	3.16	3.19	3.07	3.31	2.68
Conoco	3.45	3.60	3.51	4.03	4.81	3.64	6.78
Exxon Corp.	2.73	2.79	3.10	3.47	3.53	4.06	3.23
Kerr-McGee	7.44	6.98	7.01	6.82	7.59	NAv	NAv
USX-Marathon Group	6.38	5.84	5.66	5.61	5.83	5.28	7.24
Mobil Corp.	3.26	3.14	2.98	3.03	3.02	3.98	2.38
Phillips Petroleum	3.01	2.64	2.21	2.13	1.94	1.72	2.21
Royal Dutch/Shell	3.04	3.29	3.25	3.42	3.74	4.85	3.29
Texaco	3.06	2.85	3.02	3.04	3.23	3.63	2.41
Unocal Corp.	2.89	2.94	2.98	3.08	3.27	3.91	2.19
Average	$3.66	$3.71	$3.66	$3.79	$4.01	$3.57	$4.53

Source: Wertheim Schroder & Co., Inc.

Note: NAv = data not available.

Figure 1. Increase Required in Oil and Natural Gas Prices to Offset Potential Changes in Production Costs and Depreciation Rate
(per BOE)

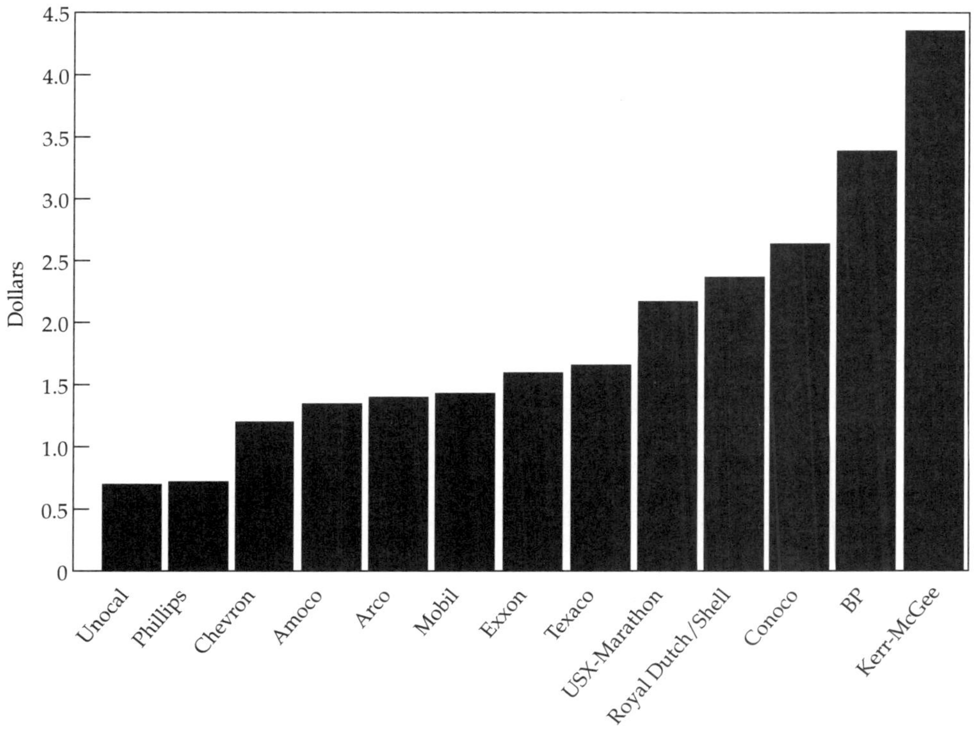

Source: Wertheim Schroder & Co., Inc., *Exploration and Production Results, 1987–91* (May 1992).

Notes: Production costs are assumed to increase 5 percent annually. Depreciation rate is the capitalized cost on the company books. $1/barrel for oil = $0.15/mcf for natural gas. Average = $1.90.

but some companies seem to be way out of line. The price of oil and natural gas would have to increase more than $3 a barrel for the earnings of BP and Kerr-McGee to remain unchanged.

Reserve Replacement Ratios

Wall Street performs a big disservice to the institutional investment community by placing undeserved emphasis on reserve replacement ratios and finding costs as measures of how well a company has done. Oil companies do not pay particular attention to these numbers, and no major oil company bases its own internal economics on them. Nevertheless, because Wall Street relies on these numbers, oil company managers find themselves constantly explaining the figures.

The industry has done a fair job with reserve replacement, as shown in **Table 5**. In the 1987–91 period, the industry replaced 113 percent of what it produced; that is, if the companies produced 100 barrels of oil in a given year, at the end of the year, they had added 113 barrels to reserves, for a net gain of 13 barrels. Based on these numbers, the oil industry is not depleting reserves. In fact, it has consistently expanded them during the latest five-year period.

Costs incurred to bring resources onto the books as proven reserves include exploration costs, acquisition costs, drilling costs, development costs, and so forth. To be counted as proven reserves, the reserves must be economically producible under current prices, current cost structures, and current tax rates. As shown in **Table 6**, the industry has done a good job of controlling these costs. Based on a five-year moving average, finding costs have fallen from $5.45 a barrel in the 1985–89 period to $4.68 in the 1987–91 period. With the help of improved technology, companies have been able to map the location of oil and natural gas reserves more accurately than in the past. They have also upgraded projects, especially in response to the oil price crash in 1986. They cut back on capital spending and funded only the projects they thought would generate 30–40 percent returns for several years.

These statistics assume that the profitability and cash flow characteristics of every barrel of oil are

Table 5. Major Oil Companies' Reserve Replacement Ratios, Including Acquisitions and Divestments—Worldwide Five-Year Averages

				Last Three Years (1989–91)		
Company	1985–89	1986–90	1987–91	Worldwide	U.S.	Non-U.S.
Amoco Corp.	131%	123%	124%	67%	44%	92%
Atlantic Richfield	111	109	106	86	72	201
British Petroleum	115	103	106	42	89	–14
Chevron Corp.	52	73	82	76	46	124
Conoco	116	114	126	113	130	99
Exxon Corp.	111	104	106	108	96	116
Kerr-McGee	94	110	148	176	131	256
USX-Marathon Group	81	84	92	101	55	208
Mobil Corp.	90	102	99	101	81	115
Phillips Petroleum	74	121	128	126	136	110
Royal Dutch/Shell	161	180	171	177	45	222
Texaco	43	59	65	66	92	15
Unocal Corp.	110	117	121	129	86	198
Average	99%	108%	113%	105%	85%	134%

Source: Wertheim Schroder & Co., Inc.

equal. They are not. Some oil is more expensive to produce than other oil; some oil sells for less; some oil is burdened by higher taxes. As a result, replacement ratios and development costs can be misused in an analysis. For example, compare Royal Dutch/Shell and Exxon, two very well-managed companies. According to the reserve replacement ratios in Table 5, Royal Dutch/Shell performed well in each of the last three five-year periods. During the most recent five-year period, it replaced 171 percent of what it produced. As shown in Table 6, to do this, it incurred finding costs of $3.13 a barrel; only Phillips had lower costs.

Exxon's reserve replacement ratio was 106 percent during the past five years, which was slightly below average. Its finding costs, at about $4.62 a barrel, were average.

Which company redeployed shareholders' capital in the best manner? According to these indicators, the answer is Royal Dutch/Shell, which found more oil at a lower cost. That answer is not necessarily true, however. What really matter are how much cash a company receives for what it produces and how that figure compares with the cost of finding and developing the reserves. As Table 2 showed, despite an excellent five-year record, Royal

Table 6. Major Oil Companies' Incurred Finding Costs (Acquisition, Exploration, and Development)—Worldwide Five-Year Average
(dollars per BOE)

				Last Three Years (1989–91)		
Company	1985–89	1986–90	1987–91	Worldwide	U.S.	Non-U.S.
Amoco Corp.	$6.03	$5.47	$5.16	$5.98	$6.07	$5.91
Atlantic Richfield	4.50	3.80	4.06	4.64	4.46	5.21
British Petroleum	7.11	6.91	6.69	8.16	3.05	17.67
Chevron Corp.	6.04	5.47	5.10	4.74	4.87	4.63
Conoco	5.62	6.02	6.18	7.22	5.61	8.99
Exxon Corp.	4.90	4.64	4.62	4.69	5.14	4.50
Kerr-McGee	8.88	6.75	5.76	5.22	5.64	4.84
USX-Marathon Group	7.17	5.75	5.59	4.87	5.88	4.01
Mobil Corp.	4.00	3.39	3.54	3.65	4.49	3.20
Phillips Petroleum	3.28	2.35	2.70	2.99	2.65	3.52
Royal Dutch/Shell	3.76	2.95	3.13	3.14	10.56	2.57
Texaco	4.68	3.95	3.95	4.07	4.13	3.99
Unocal Corp.	4.89	4.51	4.36	4.31	6.29	2.84
Average	$5.45	$4.77	$4.68	$4.90	$5.30	$5.53

Source: Wertheim Schroder & Co., Inc.

Dutch/Shell's cash flow fell from $5.94 a barrel in 1987 to $5.36 a barrel in 1991, a 10 percent drop that was slightly more than the industry average. In contrast, Exxon, which appeared to be below average when judged by the previous indicators, increased cash flow 6 percent, from $6.68 a barrel in 1987 to $7.05 a barrel in 1991. This performance was the second best in the group. Analysts who focus on cash flow would conclude Exxon did a better job than Royal Dutch/Shell.

In exploration and production, the critical variables are oil and natural gas prices and reserve replacement. Because no one can successfully predict oil and natural gas prices, management has to add reserves at a cost that is effective relative to the expected revenues from producing that oil and natural gas.

Downstream Analysis

Downstream activities include refining, marketing, and transportation. The revenues generated in this end of the business are simply the sales prices for gasoline, jet fuel, heating oil, and other products. Companies can sell products through a number of channels. An independent refiner can sell into the spot market for whatever it can get. An integrated company can sell gasoline and other products directly to its own chain of wholesalers and/or to a service station network, which will then resell it to the public. They can sell jet fuel directly to the airlines.

The key variables in this segment are the refining margin and the marketing margin. Downstream costs are essentially the costs of crude oil. For integrated companies, an increase in the price of oil benefits the upstream segment but hurts the downstream segment because of the increased cost of goods sold. The effect on the company depends on how tight the market is and how quickly higher costs can be passed along to consumers. When refinery utilization rates are low and demand is weak, companies will have difficulty passing along cost increases, and margins will suffer. When demand is robust and utilization rates are high, companies can quickly pass along cost increases.

Refining Margins

The refining margin is the difference between the cost of crude oil and the value of crude oil products on the spot market. If a company buys crude oil for $20 a barrel and sells the gasoline in the spot market for $25 a barrel, it earns a gross refining margin of $5 a barrel, not including operating expenses, distribution costs, or taxes.

Refining margins in Europe and the Far East, as exemplified by the gross margins of Rotterdam and Singapore refineries shown in **Figure 2**, have fallen sharply since the beginning of 1991. Profits have declined, and for five quarters the industry has been in a slow-growth mode in both critical foreign markets. In fact, operators of a simple refinery in the Rotterdam market have probably been losing money.

In trying to determine which companies can pass along cost increases the most quickly, analysts should examine the oil market not as one large worldwide market but as several niche markets. The situation in California, for example, is very different from that in the rest of the United States. California is like an island with two entry barriers—the Rocky Mountains and the Pacific Ocean. (Some might say a third barrier is the California Air Resources Board, which is leading the nation in setting tough environmental standards.)

Figure 3 shows recent refining margins for the Gulf and West coasts in the United States. Before 1992, the Gulf Coast refining margin tracked the West Coast refining margin. Refiners in California had typically been more profitable than refiners in the Gulf Coast, and West Coast refiners earned about 50 cents a barrel more than Gulf Coast refiners throughout 1991.

The sale of one of Royal Dutch/Shell's three West Coast refineries caused California refining margins to shoot up considerably in the December 1991–January 1992 period. Royal Dutch/Shell decided to sell its Los Angeles refinery primarily because the company was producing more gasoline than it was selling through its service stations. In addition, it did not want to make environmental expenditures at

Figure 2. Rotterdam and Singapore Gross Refinery Margins

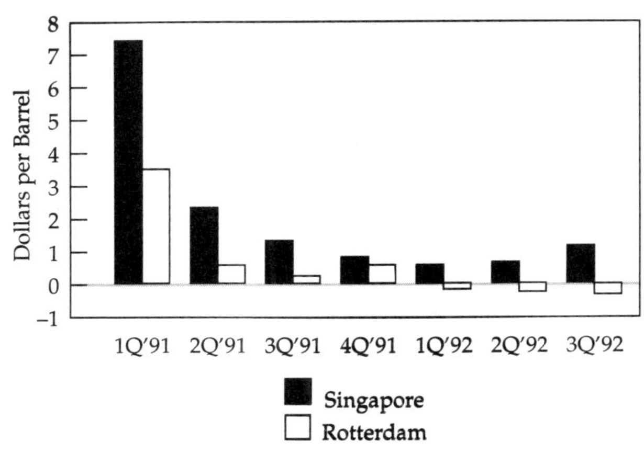

Source: Wertheim Schroder & Co., Inc.

Figure 3. U.S. Refining Margins—West Coast versus Gulf Coast

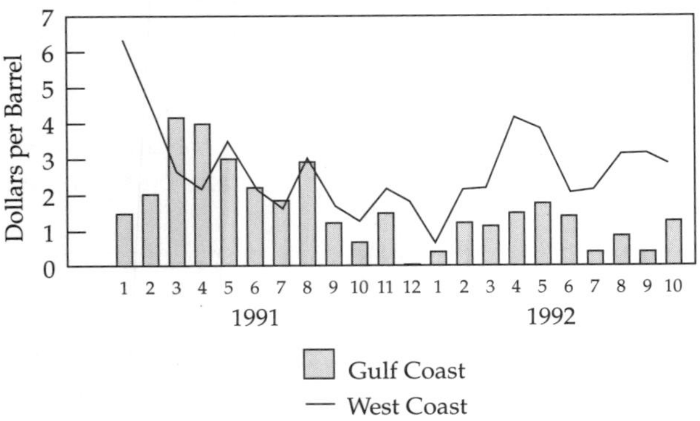

Source: Wertheim Schroder & Co., Inc.

three plants on the West Coast when it could manage with two.

Meanwhile, Unocal was undergoing a large capital expenditure program to upgrade a refinery next door to Royal Dutch/Shell's in Los Angeles. Unocal decided to buy its neighbor's refinery, join the two refineries together, and close inefficient units at both. As a result of this project, refining capacity in California fell by about 8 percent, and California refining margins immediately took off. During December and January, margins were roughly $1 a barrel; by April and May 1992, they had climbed to about $4 a barrel. The two California companies enjoyed a sustainable and systematic increase in the profitability of their remaining assets.

For many in the industry, this pattern is the promising future of refining in the United States. Refiners are hoping that the Clean Air Act will ultimately result in capacity reductions by forcing companies that are unable or unwilling to make the capital investment necessary to upgrade outdated refineries to close their plants. Those that remain will exact economic rents from a resource that will then be a scarcity rather than a surplus.

Refining profits have been systematically weak because major companies have kept pressure on spot prices. Weak refining profits have not hurt the major integrated companies, because they can make up the difference through marketing operations.

Marketing Margins

The marketing margin is the uplift an integrated company receives for selling product to its own dealers or wholesalers rather than in the spot market. The product might sell as high as $27 a barrel. Extra costs include the costs of the service station network, advertising, environmental liabilities associated with service stations, underground tanks, the cost to distribute profits to service stations, and so forth.

Figure 4 and **Figure 5** show the refining and marketing margins and the total downstream contributions for, respectively, Gulf Coast and West Coast companies that refine product and sell gasoline through their own service stations. (This analysis does not apply to all product sales, because only about 50 percent of the product from refineries is gasoline and only part of that is sold to the dealer network.) The current Gulf Coast refining margins (Figure 4) are well below those of 1991, but marketing margins have picked up, so the total downstream contribution is about equal to that of 1991. The conclusion is that an integrated company selling through its own gasoline stations is less vulnerable to falling margins than independent refiners. In addition, independent refiners may not have the cash flow and capital to make investments required under the Clean Air Act and may eventually be forced out of business. If so, pressure on refining profits will lift after 1995.

The West Coast situation, shown in Figure 5, is golden. The total downstream contribution for integrated companies that sell gasoline through service stations is currently more than $12 a barrel; in 1991, it averaged $6 a barrel. This increase in margins was achieved without an economic recovery, without an increase in employment, without an increase in the number of inhabitants, and without growth in tourism. The companies accomplished it through industry discipline. They decided that they had to recover the increased environmental costs and that the best place to do it was through marketing in those areas where distribution can be controlled. They recognized that recovering costs through refining operations would benefit the competition they were hoping to eliminate.

Table 7 shows the after-tax profits per barrel from marketing and refining, with all the special items removed, for selected companies. Although average profitability declined 10 cents a barrel between third quarter 1991 and third quarter 1992, profits for Chevron, Arco, and Unocal were up 33 cents, 62 cents, and 60 cents, respectively. All three refine and market on the West Coast. A different trend developed for other companies. Exxon, Mobil, and USX-Marathon's third-quarter earnings were burdened by maintenance expenditures, which (like exploration expenditures in the upstream) can experience uneven timing during the year. Both exploration and refinery maintenance expenditures can have a severe impact on earnings. Analysts need to adjust earnings forecasts for these items.

Figure 4. U.S. Gulf Coast Downstream Margins—Total Refining plus Gasoline Marketing

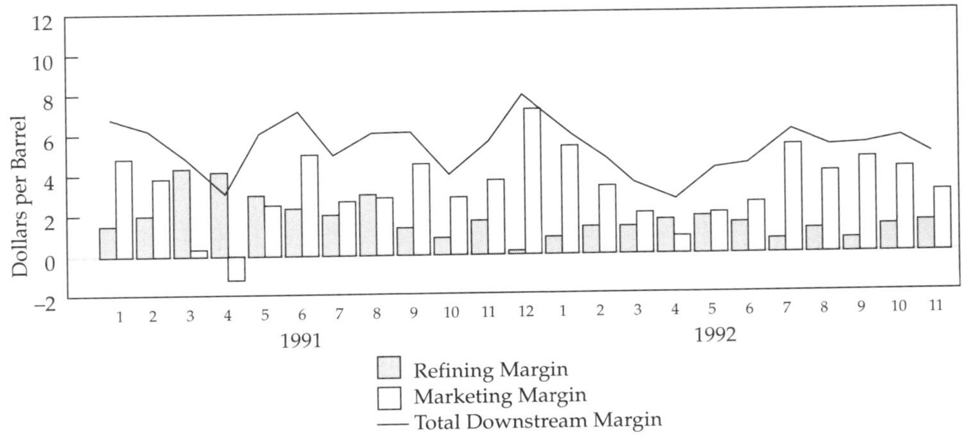

Source: Wertheim Schroder & Co., Inc.

Figure 5. U.S. West Coast Downstream Margins—Total Refining plus Gasoline Marketing

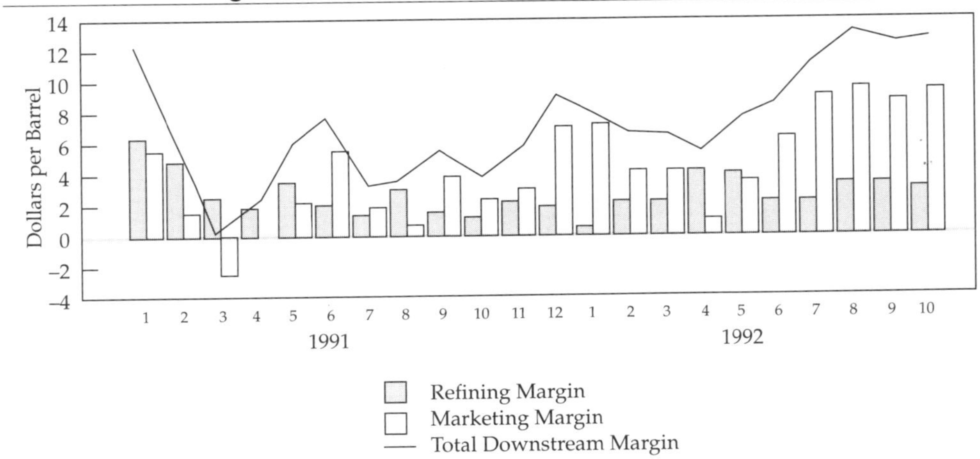

Source: Wertheim Schroder & Co., Inc.

Chemicals

Chemical earnings can have a significant impact on the total profits of the major oil companies. When forecasting earnings, cash flow, capital spending, the sanctity of the dividend, and the ability of an integrated oil company to grow, therefore, remember that a large chunk of earnings comes from chemicals, which are closely tied to the economy and may be plagued with excess capacity during the next several years.

From the peak in 1988 to the trough in 1992, chemical earnings for the major oil companies fell from almost 30 percent to approximately 7 percent of adjusted earnings. In 1988, the industry earned almost $7 billion selling petrochemicals; for 1992, approximately $1.2 billion will be earned. **Table 8** shows the trend in adjusted chemical profits as the industry expanded capacity to manufacture commodity chemicals and subsequently experienced weak growth in demand. Chemical margins mirrored adjusted profits—growing to 15 percent in 1988 before returning to the 3–5 percent range in 1991.

Other Factors

The point of speculating about oil and natural gas prices, production costs, or refining margins is to develop valid and sensible earnings, cash flow, and dividend growth estimates. Two other important factors when estimating earnings and cash flow are

Table 7. Major Oil Companies' Downstream Margins—Estimated After-Tax Profits, 1Q1991–3Q'92
(dollars per barrel sold)

								3Q'92 versus	
Company	1Q'91	2Q'91	3Q'91	4Q'91	1Q'92	2Q'92	3Q'92	2Q'92	3Q'91
Amoco Corp.	$2.05	$2.05	$1.45	$0.97	$0.58	$1.01	$1.30	$0.29	($0.15)
Chevron Corp.	1.01	0.05	0.28	(0.10)	0.30	0.45	0.61	0.16	0.33
Exxon Corp.	2.33	1.04	0.89	0.42	0.21	0.79	0.29	(0.50)	(0.60)
Mobil Corp.	1.22	0.77	0.83	(0.27)	0.01	(0.53)	0.22	0.31	(1.05)
Texaco	0.76	0.80	1.04	0.41	0.98	0.98	1.15	0.17	0.11
Arco Chemical	1.48	0.95	1.71	1.24	1.51	1.97	2.33	0.36	0.62
Phillips Petroleum	0.79	0.28	0.36	0.40	(0.06)	0.91	0.22	(0.69)	(0.14)
Unocal Corp.	(0.10)	0.17	0.41	0.52	0.19	1.01	1.01	0.00	0.60
USX-Marathon Group	1.07	1.34	0.87	0.78	0.47	0.53	0.23	(0.30)	(0.64)
Average	$1.18	$0.83	$0.87	$0.49	$0.47	$0.79	$0.77	($0.02)	($0.10)

Source: Wertheim Schroder & Co., Inc.

Table 8. Major Oil Companies' Chemical Profits, Adjusted, 1985–93[a]
(millions of dollars)

Company	1985	1986	1987	1988	1989	1990	1991	1992[a]	1993[a]
Amoco Corp.	$ 195	$ 264	$ 469	$ 727	$ 584	$ 240	$ 93	$ 201	$ 300
Chevron Corp.	27	174	231	455	375	192	117	46	80
Exxon Corp.	147	470	680	1,363	1,082	552	512	500	600
Mobil Corp.	53	113	285	597	558	339	217	87	135
Texaco	10	51	81	246	268	48	23	0	25
Royal Dutch/Shell	230	830	1,054	1,825	1,611	836	156	26	250
Arco Chemical–Chemical	107	132	260	400	336	284	212	261	260
Arco Chemical–Lyondell[b]	(25)	132	137	529	214	223	111	14	80
Phillips Petroleum	219	299	230	595	431	322	194	47	100
Unocal Corp.	58	38	27	57	47	41	50	27	40
Total	$1,021	$2,503	$3,454	$6,794	$5,506	$3,077	$1,685	$1,209	$1,807

Source: Wertheim Schroder & Co., Inc.

[a]Estimate.

[b]Arco Chemical sold 54 percent of Lyondell in an initial public offering in January 1989 and in 1992 owned 50 percent.

Table 9. Major Oil Companies' Debt-to-Capitalization Ratios, 1985–93[a]

Company	1985	1986	1987	1988	1989	1990	1991	1992[a]	1993[a]
Amoco Corp.	27%	26%	24%	33%	32%	30%	29%	32%	32%
Chevron Corp.	38	39	34	33	35	31	34	38	38
Exxon Corp.	21	20	19	23	35	29	27	28	28
Mobil Corp.	45	41	37	32	30	30	32	34	35
Texaco	43	44	55	49	40	39	40	41	42
Royal Dutch/Shell	25	21	19	17	16	17	19	20	22
Arco Chemical	58	62	60	52	50	51	55	55	55
Phillips Petroleum	81	78	78	70	65	59	59	61	60
Unocal Corp.	78	76	74	68	64	62	66	63	59
USX-Marathon Group	NAv	NAv	NAv	58	55	53	58	54	54
Average	46%	45%	44%	44%	42%	40%	42%	42%	42%

Source: Wertheim Schroder & Co., Inc.

Note: NAv = data not available.

[a]Estimate.

the debt-to-capitalization ratio and the dividend as a percentage of cash flow. **Table 9** presents debt-to-capitalization ratios for the major companies. It shows that, during the past five years, the companies have done well in protecting their balance sheets. In 1985, when oil was $27 a barrel, the average debt-to-capitalization ratio in the industry was 46 percent. In 1992, after six years of generally declining energy prices and poor U.S. natural gas prices, the average debt-to-capitalization ratio in the industry was an estimated 42 percent. The ratio has improved despite the considerable tightening of accounting standards defining equity. The improvement would have been more dramatic if the current balance sheets had been recast according to accounting standards prevailing in 1985.

The oil industry has the flexibility to defer major capital projects without cutting into current earning power. Oil prices crashed in 1986 and have remained depressed. In response to this environment, companies cut capital expenditures from $7.8 billion in 1985 to about $4 billion in 1987. Capital expenditures reached the $7 billion level again only in 1991 and 1992. Moreover, today's $7 billion will buy more than in the 1980s because of reduced costs, improved technology, and the better terms being offered by foreign countries that were previously closed to major U.S. oil companies.

The picture for dividend payout ratios, shown in **Table 10**, is less rosy. The average ratio in the mid-1980s was about 20 percent, but it has increased to about 30 percent in recent years. Dividend growth during the next couple of years will be lower than the historical rate.

Valuation

Earnings and cash flow models are appropriate valuation tools for the oil industry, but the models work best when used together. Exxon, which is involved in all phases of the business, provides a good example of why earnings models alone can fall short. In 1986, when oil was $16 a barrel and natural gas was $1.63 an mcf, Exxon earned $3.45 a share, adjusted for special items. In 1992, with oil at $20 a barrel and natural gas almost unchanged at $1.65, Exxon should earn $3.55 on an adjusted basis. Despite a $4-a-barrel increase in oil prices, earnings per share are virtually unchanged. The company made more money in the upstream as oil prices rose (exploration and production earnings increased from $698 million in 1986 to $954 million in 1992) but saw downstream earnings (refining and marketing and chemicals) fall by a similar amount.

Based on current earnings and cash flow valuations, oil stocks as a group appear to be fairly, perhaps even fully, valued. **Figure 6** shows P/E multiples relative to the S&P Industrials. Based on 1993 earnings estimates—and assuming $20-a-barrel oil prices, $1.80 an mcf for natural gas, and some modest improvement in refining and chemicals—the group is currently trading at about 86 percent of the market multiple, which appears high by historical standards. Contributions from chemicals and refining are at trough levels, but refining earnings should improve if capacity is reduced as a result of the Clean Air Act. As shown in **Figure 7**, at a 70 percent relative cash flow multiple, the group trades at the high end of the range relative to the S&P Industrials.

On a yield basis, seen in **Figure 8**, a completely different signal muddies the waters: The group is

Figure 6. Major Oil Stocks' Relative P/E Multiples versus Posted WTI Oil Prices

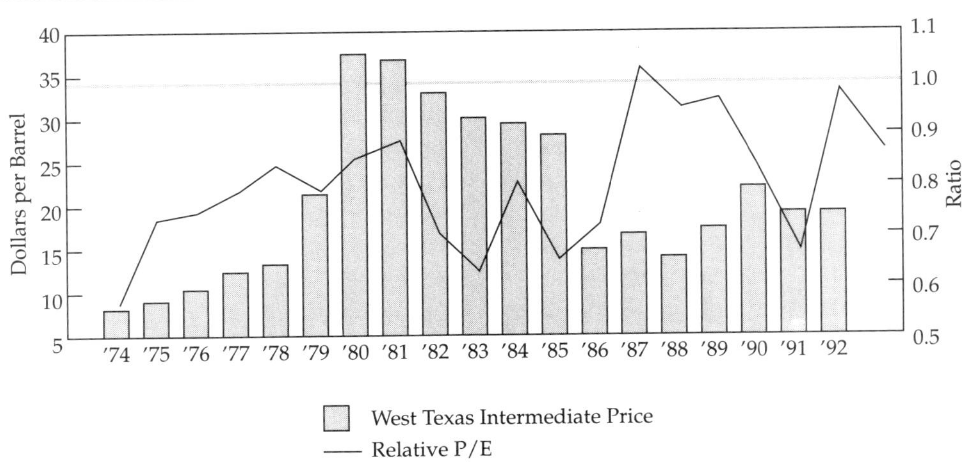

Source: Wertheim Schroder & Co., Inc.

Figure 7. Major Oil Stocks Relative Price-to-Cash-Flow Multiples versus Posted WTI Oil Prices

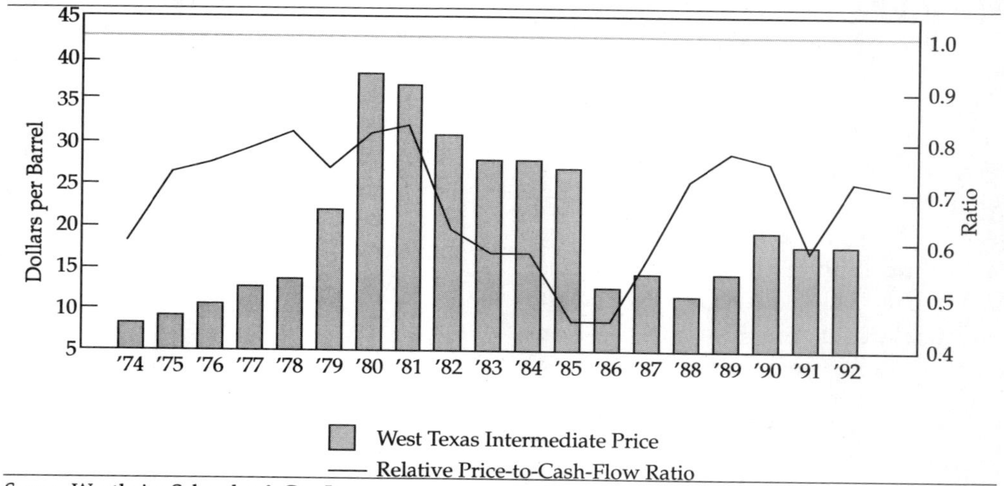

Source: Wertheim Schroder & Co., Inc.

selling at about 173 percent of the market, which is extremely cheap. This relative yield is actually higher than the yield in 1986, when the price of oil was $10 a barrel and some companies cut their dividend. According to the yield parameter, the market is viewing the oil stocks more negatively today, relative to other investment alternatives, than it did in 1986.

Figure 8 also indicates the changes in the market's expectations for oil stocks and the value the market was willing to place on those expectations. The 1974–80 period was characterized by sharply rising oil prices, rising demand, stable cost structures, and expectations for rising oil prices. The relative yield for the group decreased from 150 percent of the market to only 100 percent, a 50 percent improvement in relative valuation. Oil stocks peaked in 1980–81, coincident with the peak in valuation as shown by the relative yield. The industry had peak oil prices and peak oil price expectations, and companies were rolling in money.

Valuation collapsed from 1980 to 1985. The quoted price of oil fell from $40 to $27 a barrel, but in light of the inflation of the period, the real price drop was even more dramatic. Demand collapsed, no money was being made in commodity chemicals or refining, and companies were leveraging up their balance sheets. Chevron and Texaco, with 12–13 percent debt-to-capital ratios, leveraged up to 50 percent to make big acquisitions. Other companies tried to diversify into retailing, mining, computer systems, and other unrelated businesses. The stock market paid companies for this diversification with a 60 percent decrease in value relative to the S&P Indus-

Figure 8. Major Oil Stocks' Relative Yields, Inverted, versus Posted WTI Prices

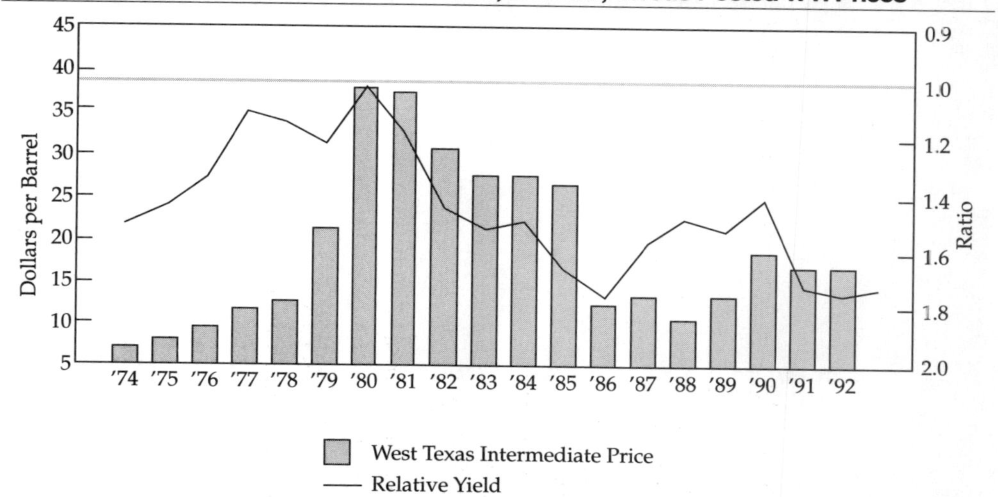

Source: Wertheim Schroder & Co., Inc.

Table 10. Major Oil Companies' Dividend Payout Ratios as a Percentage of Cash Flow from Operations, 1985–93[a]

Company	1985	1986	1987	1988	1989	1990	1991	1992[a]	1993[a]
Amoco Corp.	19%	25%	21%	20%	24%	23%	34%	36%	29%
Chevron Corp.	16	28	24	22	32	23	33	32	28
Exxon Corp.	25	32	27	29	38	31	31	36	35
Mobil Corp.	22	26	26	27	25	29	33	38	34
Texaco	15	17	6	17	107	30	33	38	36
Royal Dutch/Shell	15	24	26	28	24	27	33	37	35
Arco Chemical	19	25	20	20	24	26	32	30	30
Phillips Petroleum	29	18	16	10	19	17	31	29	25
Unocal Corp.	13	10	9	11	12	15	17	17	14
USX-Marathon Group	NAv	NAv	NAv	27	29	28	28	29	17
Average	19%	23%	20%	21%	34%	25%	30%	32%	28%

Source: Wertheim Schroder & Co., Inc.

Note: NAv = data not available.

[a]Estimate.

trials. Yield on the stocks rose from about 100 percent of the market to about 160 percent.

Analysts attempt to pull all these numbers together and come up with a bottom line: Should investors buy these stocks? If so, which ones? The critical factors that analysts should consider are the price of oil, the price of natural gas, worldwide downstream margins, and expected chemical margins. From forecasts of these prices and margins, a model will yield expected dividends, earnings, and cash flows. The most useful model will allow an analyst to examine the effect on earnings of changes in the critical factors.

Table 11 uses relative P/E, cash flow, and yield together to determine the value of various stocks. According to normalized valuations based on WTI at $21 a barrel and U.S. natural gas at $1.80 an mcf, the group should trade about 10 percent higher than it is trading in late 1992. The market does not expect oil prices to increase, however; it is worried about what will happen as Kuwait continues to increase its oil production and when Iraq inevitably resumes production. The market is not forecasting $21-a-barrel oil; it is focusing on the potential downside risk.

Table 11 contains a trading range for the group based on $17-a-barrel oil on the downside and $23-a-barrel oil on the high side. These companies have never traded at a discounted valuation based on less than $17-a-barrel oil during the past six years, even though oil was $13 and $14 a barrel several times. This stability rests on investors' beliefs that the dividend yield for most of these companies is safe. In addition, even though oil was $30–$40 a barrel during the two weeks after Iraq invaded Kuwait, the market never discounted more than $25 a barrel. Thus, the downside risk in these stocks is about 7 percent at the depths of pessimism, which was in February or March of 1992; the upside potential is 21 percent at the height of optimism.

Table 11. Major Integrated Oil Stocks—Valuation Ranges and Target Prices

		Target Price Based on Potential Trading Range				Normalized Valuation				
						Based on Multiples of				
Company	Recent Price	WTI = $17/bbl Natural Gas = $1.50/mcf	WTI = $23/bbl Natural Gas = $2.00/mcf	Percent Risk	Percent Reward	Earnings per Share	Cash Flow	Yield	Target Price[a]	Percent Change
Amoco Corp.	$ 51	$ 44	$ 56	–14%	10%	$ 53	$ 51	$ 51	$ 51	0%
Chevron Corp.	68	64	83	–6	22	76	80	71	76	11
Exxon Corp.	59	55	66	–6	12	64	58	63	62	5
Mobil Corp.	60	59	75	–1	25	66	69	71	69	14
Texaco	60	55	74	–8	23	64	68	67	66	10
British Petroleum (BP)	44	NAv	NAv	NAv	NAv	NAv	NAv	NAv	NAv	NAv
Royal Dutch	82	82	98	0	19	88	93	92	91	11
Shell	50	44	54	–12	8	48	47	53	50	–1
Arco Chemical	111	98	134	–12	21	116	123	119	120	8
Phillips Petroleum	24	22	30	–7	27	26	29	27	27	14
Unocal Corp.	25	23	34	–9	35	29	35	26	30	19
USX-Marathon Group	17	17	22	–3	29	16	27	16	20	16
S&P Industrials	491	491	491	0	0					
Average without BP				–7%	21%					10%

Source: Wertheim Schroder & Co., Inc.

Note: NAv = data not available.

[a]Based on weighting of 33 percent earnings-per-share valuation, 33 percent cash flow valuation, and 34 percent yield valuation.

Question and Answer Session

Michael L. Mayer

Question: Is there a time-lag issue in the Exxon–Shell comparison? That is, will low incurred cost and high reserve replacement costs at Shell not have much impact on cash flow for the current period but benefit cash flow in future periods?

Mayer: One way to normalize that effect is to use five-year moving averages. In this presentation, I have looked at a total of 8 years. Most major projects in the industry take 3–5 years from the time they are discovered to the time they are producing oil, although some projects take 10–12 years. Arco is working on natural gas projects in Indonesia and China, where they found natural gas about 15 years ago and incurred the exploration expense in that period. Of course, none of that expense is in these numbers. Analysts must deal with the data they have and do some smoothing. We do not have good data from all these companies going back 15 years.

Royal Dutch/Shell purchased a lot of reserves in Nigeria, which made its numbers look good. In Nigeria, the guaranteed profit margin is approximately $2.25 a barrel. If somebody offers $2.25 a barrel every year for the next 15 years, an investor will pay the discounted present value for that earnings stream, or about $1.25 on a per-barrel basis. That is exactly what Royal Dutch/Shell paid for those reserves—about $1.25 a barrel. The company included this purchase in its finding costs at $1.25 a barrel, and that is exactly what it is worth.

Question: Oil companies are valued on cash flow, so why does it matter if future DD&A expenses drift upward?

Mayer: Earnings matter. Analysts insist the real indicator is cash flow, so that is what they worry about, but if a company has a disappointing quarter, the stock will drop $1.50 or $2 a share in a single day. Also, the numbers tend to creep upward. Analysts begin with the assumption that oil prices will increase 3 percent a year, a little less than $1 a barrel. The price must increase by this amount for the industry to break even on its existing production and recover increased depreciation charges and increased production costs. If the analysts figure prices will increase and then calculate the earnings impact but ignore changes in depreciation rates for the industry, they will be way off on earnings. They will have a much more optimistic outlook for the earnings, cash flow, and ultimate dividend growth for that industry than the data warrant.

More importantly, analysts who do not distinguish among companies when they buy oil stocks will probably get burned. For example, Phillips and Unocal seem to be well positioned. They could have positive earnings surprises as their depreciation rates drift down, but British Petroleum and Kerr-McGee could have negative earnings surprises as their depreciation rates increase.

Question: Please describe alternative accounting approaches for exploration and production activities and how they affect the numbers.

Mayer: The companies I follow use successful-efforts accounting, which differs from the full-cost accounting the small companies tend to use. With successful-efforts accounting, companies that find oil capitalize the costs and charge it against earnings on a unit-production basis. If they do not find oil, they charge it against earnings immediately.

Question: How have the managers of oil companies reacted to your analyses of their efficiency? Have your results been challenged?

Mayer: The response has been mixed. Chevron, which came out in the middle of the pack, invited me to give a presentation to its exploration and production people. Phillips, which ranked third, requested 200 copies to distribute to its staff. With British Petroleum, which came out on the bottom two years in a row, I had revealed the results to company management in advance and asked if I had made any mistakes or had any intellectual or conceptual problems. They told me I hadn't, and a few weeks later, they took some big write-offs and cut the dividend in half.

Question: Where will West Coast margins go now that the California price war is over and Arco has ended its long-standing trend of lowering price?

Mayer: Let me provide some background on the California market. Prior to Saddam Hussein's invasion of Kuwait, Arco priced its gasoline 3–5 cents below its competitors' prices. It wanted every barrel of crude oil it produced in Alaska and California to go through its two refineries on the West Coast, and every barrel of that to go through its own service stations. That strategy was very successful. Arco made a lot of money selling not only gasoline but also convenience items. The stations were all self-serve—pump your own gas and buy a hot dog and drink.

Arco will not acknowledge this change, but after Hussein invaded Kuwait, it changed its pricing strategy relative to other companies. Its prices are now about the same as the competition in many markets. Relative to the industry, it has improved its gasoline price by 4 or 5 cents a gallon, which has very powerful earnings implications. Sensitivity analysis indicates that every penny-a-gallon improvement in Arco's downstream margins is worth 30 cents a share in earnings. The company currently earns about $6 or $7 a share. If it can improve its relative gasoline margin by 4 cents, it adds $1.20 to earnings. That is equivalent to the price of oil increasing about $1.50 a barrel.

Question: Does the industry have any common or standardized method for transfer pricing between upstream and downstream operations? If not, how can one compare the performance of fully integrated and independent companies when barrels of oil differ in quality?

Mayer: Every company says it has arms-length transactions between its upstream and downstream segments: The production unit sells the oil to the refining unit at market value.

Interestingly, Arco's downstream margin and differential performance with respect to competitors have been the best in the business for the past 10 years, and some companies have charged that Arco underprices its Alaskan oil and sells it into the California and Washington state refinery systems at below-market value. Arco has had an interest in keeping Alaskan prices low, because Alaska imposes a high severance tax and a royalty tax equaling about 30 percent of the value of the oil. For every dollar understatement, Arco could save 30 cents and transfer that to the refinery. The market value of Alaskan oil is hard to determine, however, because a couple of companies produce and use it all, so the market does not have much price transparency. Alaskan state authorities have sued Arco on this issue, and Arco has made settlements of hundreds of millions of dollars, which have, allegedly, cleared up the problem.

Actually, oil companies do not sell oil; the companies analyzed in this presentation are net buyers of oil. They sell gasoline. Their profits depend on economic growth and their ability to sell the gasoline products at a fair price that recovers not only crude oil costs but also prospective environmental spending.

Valuing Oil Securities

Thomas P. Moore, Jr., CFA
Senior Vice President and Director of Equity Research
State Street Research & Management Company, Inc.

Valuing oil securities requires familiarity with such industry practices as use of barrel-of-oil equivalents and knowledge of the difference between Canadian and U.S. reserves. Given such caveats, three valuation techniques predominate: asset valuation, ratio of price to cash flow, and relative dividend yield.

The oil industry differs from most other industries in at least one critical aspect: It is interwoven with world politics. Because of oil, political turmoil in the Middle East has a direct effect on everyday life in the United States and elsewhere. Governments understand that oil is vital to the security of their nations and is a key to world stability. Even though oil plays this critical world role, however, it is not a growth business. It is a cyclical industry with little earnings predictability.

Valuation of Companies

Valuation in the oil industry presents certain challenges. The industry includes a wide spectrum of companies. Oil analysts follow fully integrated international companies (such as Exxon) with multi-billion-dollar market capitalizations and operations in every part of the business; they also follow companies specializing in specific aspects of the business (such as producing or refining) with market capitalizations of less than $50 million. Valuation characteristics differ among these companies. The price of Exxon stock may react only to a major move in oil prices, yet a single well may have a significant, overnight impact on the stock of a company such as Devon Energy.

Massive amounts of data are available to oil analysts. In 1969, a company's reserve data were extremely difficult to obtain. In fact, in 1969, companies may not have even known their reserves. If they did, they were not likely to disclose the information, which was restricted to a limited number of people within the company. In the 1990s, most companies provide full disclosure, presenting data in industry supplements and sometimes on computer diskettes.

Such abundance of data presents industry analysts with a challenge: how to synthesize the data into relevant information. Relevance is the key to valuation. Knowing everything about the industry is not beneficial unless analysts can synthesize the data for portfolio managers and focus on the key elements driving a particular stock. Valuations must have solid basics behind them. It is important both to know the numbers and to understand the assumptions behind the numbers before drawing conclusions and making investment recommendations.

Valuation requires accurate statistics. Today's computerized statistical data bases offer tremendous advantages over past information sources, but the analyst still adds value to the presentation of data. When examining statistics, analysts must recognize relevant items and analyze them thoroughly. For example, companies report earnings net of nonrecurring items. These reported earnings must be adjusted to be of value to an analyst. Each slot on a spreadsheet contains only a single number, but many numbers were analyzed to get to that one number.

Valuing securities in the oil industry requires familiarity with certain industry practices. The first is the use of barrel-of-oil equivalents (BOEs). Generally, oil-equivalent barrels are calculated so that oil and gas reserves are presented on a comparable basis. One barrel of oil contains six times more Btus (British thermal units) than does a thousand cubic feet of gas. The problem with using oil-equivalent barrels is that the figures usually do not reflect the economic value of the oil or gas. For example, if a company's BOE reserves are 90 percent gas, and gas is selling for $1.10 a thousand cubic feet (mcf), the oil-equivalent barrel is worth about $6.60, even though oil is selling for $18 a barrel. Thus, the company is realizing in the marketplace only a third of the value the shareholders think is shown on the balance sheet. Analysts should use caution and judgment with barrel-of-oil equivalents and should determine the number and location of the oil barrels as well as the location and valuation of the gas. Gas outside the United States is priced much closer to the value of oil than is U.S. gas.

A second industry norm is the difference between Canadian and U.S. reserves. Canadian companies' annual reports reflect gross reserves, which do not include royalty payments. To equate Canadian and U.S. reserves, analysts must know the royalty rates for Canadian producers. The difference between gross and net reserves can range between 12.5 and 25 percent.

Analysts valuing oil companies should also know who is estimating reserve data—the company's internal reservoir engineer or an external reservoir engineer. Frequently, the numbers are virtually the same, but in some cases, an internal reservoir engineer's job may depend on pleasing the company president with the numbers; an external estimator may be more objective.

Given these industry caveats, different valuation alternatives are available:

- Ratio of price to cash flow.
- Ratio of price to free cash flow.
- Ratio of price to liquidation value.
- Ratio of price to going-concern value.
- Ratio of price to leveraged-buyout value.
- Ratio of price to earnings before interest, taxes, and depreciation.
- Ratio of price to SEC book value.
- Cash flow multiple adjusted for net debt and reserve life.
- Cash flow return on book assets.
- Cash flow return on estimated liquidation value.
- Sensitivity to changes in oil and gas prices.
- Growth rate of earnings.
- Return on equity.
- Appraised net worth.

Each measure is meaningful for a particular company. An analyst's job is to use the right one for the right company at the right time.

Three Approaches to Valuation

For large, integrated oil companies, three approaches to valuation are most relevant: asset value, ratio of price to cash flow, and relative dividend yield. These measures may not be appropriate for small exploration companies, however.

Asset Value

With the asset value approach, analysts assign a current value for reserves, refining, marketing, and other assets and then deduct debt and other liabilities to determine a company's net asset value. To assign a value for reserves, the stream of future earnings is discounted at an assumed production rate over the life of the reserves. Using a rate consistent with today's value yields $4 to $6 a barrel as the value of reserves.

The same approach applies for natural gas, but the value often depends on where the natural gas is located. Today, access to pipelines is a key factor in valuation of gas reserves. If natural gas is in the Rocky Mountains and is difficult to access, it will have less value than if it is in the Gulf of Mexico, where gas is readily shipped into the pipeline. Canada has plenty of gas but very few pipelines, which leads to lower reserve values. Several factors affect valuation of prospective acreage: How will the acreage be developed? What is the time frame for development? Does the company have sufficient financial capability to develop the acreage? In good times, the value for prospective acreage seems to increase dramatically, but one poor period can prove that prospective acreage actually has little value. For example, in the early 1970s, one company sold 1,200 of about 660,000 acres held in the Canadian Arctic. The Canadian buyer sold it for about $2,000 an acre. The original owner then valued all the remaining acreage at the $2,000 figure. For nine months, the owner's stock was one of the darlings of the market. When the market for energy stocks became tighter in 1971 and 1972, however, this company experienced one of the first debacles.

The next step in the valuation process is establishing the value of refining and marketing assets. Analysts painstakingly develop valuation parameters for the oil and gas side of an integrated company but then rather haphazardly assign an arbitrary single number (say, $5,000 a daily barrel) to the refining assets. Refineries are not all alike. The location of the refinery, how well it was built, and how well it meets environmental standards must be considered. For marketing assets, analysts should look at the value of the products' market shares. A company with 1 percent market share in a given gasoline or home-heating-oil market cannot value its marketing assets at the same price as a company with a 7–8 percent market share. Analysts must add perspective. No single answer exists for valuing these assets.

Other assets of integrated oil companies must

also be valued. Typically, analysts combine into a one-line item the valuations of other assets, including coal, geothermal, or other nonrelated companies. Valuing coal assets presents a dilemma because of their tremendous reserves but low profitability. Unless the coal is profitable at the current market price, assigning any value to coal assets is questionable. Geothermal energy carries the same problem. Geothermal was praised as a growth business in the mid- to late 1970s. Today, however, the business is no longer growing and is barely profitable; thus, assigning a positive value to these assets is difficult unless specific contracts make a particular business profitable.

These individual values lead to an overall net asset value for the company. **Figure 1** shows that exploration and production companies are currently trading near historical lows with respect to their net asset values. These lows were reached twice before in the recent past, in 1979 and 1983. In each case, positive stock price performance followed—for the entire industry in 1979 and for those severely undervalued, asset-rich companies that were specific merger targets from 1983 to 1985, such as Cities Service, Gulf Oil, and Superior Oil.

Figure 1. Ratio of Stock Prices to Net Asset Values—Exploration and Production Companies, 1978–3Q'92

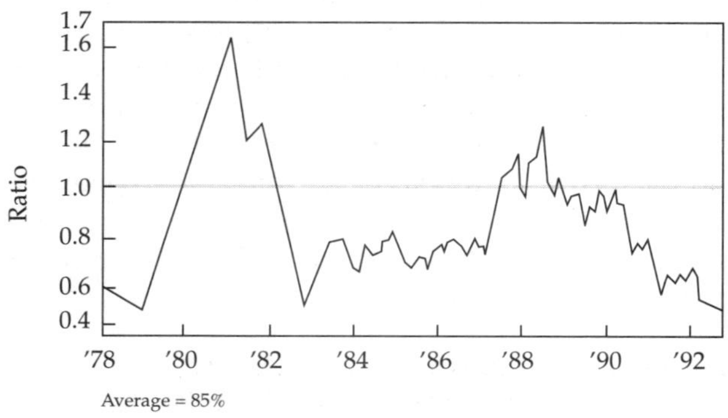

Source: PaineWebber.

Using the net asset value approach to stock picking among integrated oil companies can be hazardous, particularly in a volatile environment. Under the net asset value approach, the idea is to sell companies trading at a high percentage of asset value and buy companies trading at a low percentage of asset value. An example of this tactic's results is presented in **Table 1**. In 1980, an investor might have sold Arco Chemical, trading at 80 percent of its asset value, at $50 a share, and bought Texaco, trading at only 38 percent of its asset value, at $38 a share. By 1985, however, an Arco share had risen to $57, a 13 percent increase from its 1980 price. During that same period, Texaco shares declined almost 6 percent, although the market showed a healthy 58 percent increase.

Table 1. Asset Values versus Average Stock Prices, Selected Oil Companies, 1980, 1985, and 1990

Company	Asset Value	Average Stock Price	Stock Price as a Percent of Asset Value
1980			
Amoco Corp.	$ 60.88	$ 31.00	51%
Arco	62.95	50.53	80
Exxon Corp.	39.46	17.25	44
Phillips Petroleum	39.18	16.08	41
Texaco	129.15	38.07	29
Unocal Corp.	44.60	16.39	37
S&P 500		118.68	
1985			
Amoco Corp.	58.35	31.42	54
Arco	96.85	57.19	59
Exxon Corp.	39.50	25.80	65
Phillips Petroleum	39.20	13.34	34
Texaco	88.95	35.89	40
Unocal Corp.	37.40	17.63	47
S&P 500		187.30	
1990			
Amoco Corp.	76.00	53.71	71
Arco	159.35	121.41	76
Exxon Corp.	56.10	48.73	87
Phillips Petroleum	60.60	26.23	43
Texaco	78.60	59.50	76
Unocal Corp.	48.15	29.55	61
S&P 500		334.27	
Assumptions	*1980*	*1985*	*1990*
Discount rate	10%	15%	12.5%
Crude oil price[a]	$100	$75	$38
Natural gas price[b]	16	9	5

[a]Per barrel, for the year 2000.
[b]Per mcf, for the year 2000.

In 1985, the asset value approach gave a buy signal for Phillips Petroleum at 34 percent of valuation and a weak sell signal for Exxon at 65 percent. During the next five years, Phillips increased 96 percent in value while the marketplace increased only 79 percent. Exxon stock increased in value by 92 percent during the same period. Obviously, taking this approach to investing in the early 1980s did not help portfolio managers or analysts beat the market. During this period, however, many oil companies selling at low asset values became takeover targets, which brought investors a handsome return.

A major reason that the net asset value approach went astray is that valid assumptions are difficult to formulate for this volatile industry, especially during uncertain periods. In 1980, the discount rate assumption was 10 percent, and price expectations for the year 2000 were $100 a barrel for crude oil and $16 an mcf for natural gas. At the time, oil had risen to $38 a barrel, and most analysts expected oil to hit $100 a barrel in the next 10 years. In that environment, envisioning oil at $100 a barrel by the year 2000 was conservative. Times change! In 1985, oil was about $28 a barrel, the discount rate was 15 percent, and price expectations for the year 2000 were $75 a barrel for crude oil and $9 an mcf for natural gas. Some of the new gas that had come on-line in the late 1970s and early 1980s had actually hit $9 an mcf, so this expectation was not unrealistic.

Today's valuations are much more conservative. Analysts have been chastened during the past 12 years, and expectations are lower. Today's discount rate is 12.5 percent. Price expectations for the year 2000 are $38 a barrel for crude oil and $5 an mcf for natural gas. The moral here is that analysts should be careful with assumptions when using asset value per share, because assumptions are no more than a best estimate at a given time. Analysts and investors can only hope that today's assumptions provide a realistic estimate of the companies' future stock values.

Ratio of Price to Cash Flow

Cash flow is the key to the operations of any company. The most generic definition of cash flow begins with net income, then adds back depreciation, depletion, amortization, the after-tax impact of exploration expenses, and deferred taxes. Because specific definitions vary, analysts must be clear on the exact calculation when comparing companies on the basis of price-to-cash flow ratios.

On a long-term basis, cash flow is not as volatile as earnings, as **Figure 2** shows. Unless a substantial write-off causes depreciation, depletion, and amortization to change dramatically, future cash flows are easier to estimate than future earnings.

As **Figure 3** illustrates, cash flow per share does not follow the price of oil closely. In fact, the price of oil appears to be one of the more irrelevant factors with respect to valuation of oil companies. Cash flow analysis, on the other hand, can be helpful over time. **Figure 4** shows Chevron, Exxon, and Texaco trading at similar cash flow multiples in the early 1980s. In 1984, Exxon broke out of this group, and its price-to-cash-flow multiple through the early 1990s exceeded those of the other companies. Someone investing on this basis might conclude either that Exxon is overpriced compared with the others or that some other factor—perhaps the quality of its earnings—is making investors pay a higher multiple of cash flow for Exxon's earnings. Exxon stock established a strong trend in 1984; when the others declined in 1985 and 1986, it stayed up. Over that period, Exxon was 25–30 percent ahead of the other oil stocks; it was also ahead of the market.

Figure 2. Cash Flow per Share and Earnings per Share—Amoco, December 1975 to December 1991

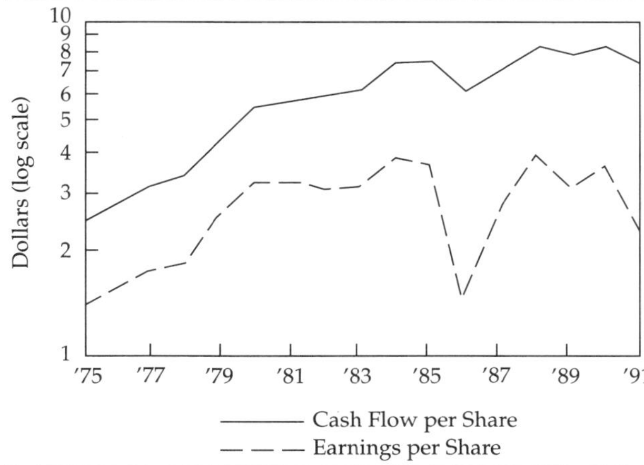

Source: Compustat data through Factset.

On a long-term basis, dramatic moves in cash flow can be used to assist the valuation decision. Therefore, analysts should examine cash flow over time to gain perspective on relative values. The advantages of cash flow are that it is more stable than net income and that company operating managers generally manage to a cash flow budget. Cash flow is easy to talk about with management, because both analysts and operating managers understand the concept. Cash flow allows analysts to evaluate the operations of a company the same way company managers do.

Along with these advantages, however, cash flow analysis also has some disadvantages. Compa-

Figure 3. Oil Prices and Cash Flow per Share—International Oils, 1980–92[a]

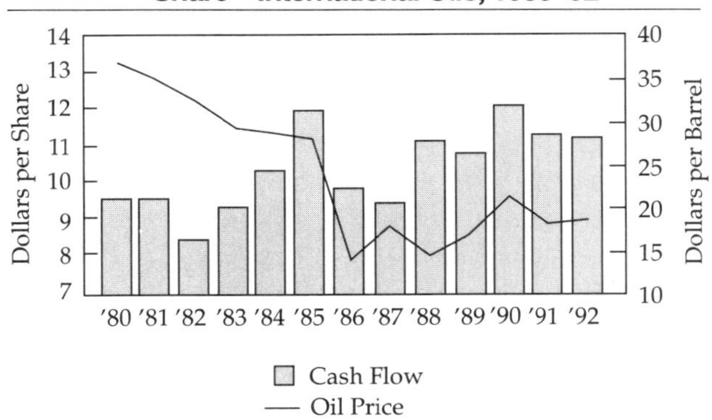

Source: Merrill Lynch & Co., Inc., "International Oils—Assessment of Trends and Outlook" (October 1992).

[a]Estimate.

Figure 4. Cash Flow Multiples—Exxon, Chevron, and Texaco, 1980–3Q'92

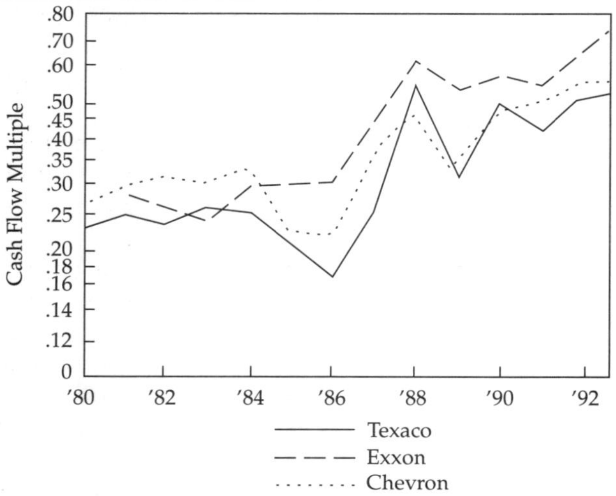

Source: Compustat data through Factset.

nies in the oil industry tend to trade within a very tight grouping around cash flow multiples. Cash flow can also be misleading: Companies may have high cash flows because of high depreciation of aging or wasting assets; thus, the future earning power of their cash flow may be impaired.

Relative Dividend Yield

Tony Spare introduced the relative dividend yield approach to valuation in his book *Relative Dividend Yield*.[1] This analytical approach is very helpful for valuing major oil companies. Most large U.S. oil companies pay above-market dividends, and they seldom, if ever, cut the dividend. As a result, their stocks are amenable to the relative dividend yield technique.

The relative dividend yield is calculated by dividing the annual dividend rate by, first, the stock price and, next, by the yield of the market index. The result is then charted over time to establish a framework of buys and sells. The challenge for analysts using this approach is to find the right spread for each company.

Figure 5 plots the relative dividend yield of Chevron from 1962 through 1991. In this example, the framework would suggest buying when the relative yield is 160 to 200 percent of the S&P 500 and selling when it is 80 to 100 percent. This process would have led to buying Chevron during the 1970–75 period and selling when the yield dropped below 100 percent in 1980, thus avoiding the pain of the 1980–85 period.

The technique would also have worked well for Mobil and Exxon, which moved in and out of established trading ranges between 1970 and 1992. The same technique would have kept an investor out of Merck during the 1970s and 1980s, however, when it was an excellent stock to own. Analysts should thus be selective when choosing companies to value with the relative dividend yield approach. The technique serves best as a check on other valuation methods, and it works best for stocks with above-market yields.

The relative dividend yield method has three advantages. First, it requires little guesswork; dividend rates are usually stable. Second, it does not require estimates of future earnings or margins, as does a dividend discount model. Third, a stock-price-to-dividend comparison reflects the market's concern for the stock itself, not necessarily for the company's operating qualities. The disadvantages to the approach are that it requires a long-term view; it misses most small-capitalization companies, which pay little or no dividend; and it is not very responsive to growth stocks.

Profitable Oil Investing

The keys to successful long-term investing in the oil industry are communication with company management, dividend reinvestment, and a long-term commitment to investing.

Communication with company management. Analysts contribute to the investment process through their ability to synthesize management views on company objectives and the analysts' assessments of management's ability to achieve ambitious financial goals. Analysts must determine whether management's shareholder goals corre-

Figure 5. Chevron Dividend Yield Relative to the S&P 500

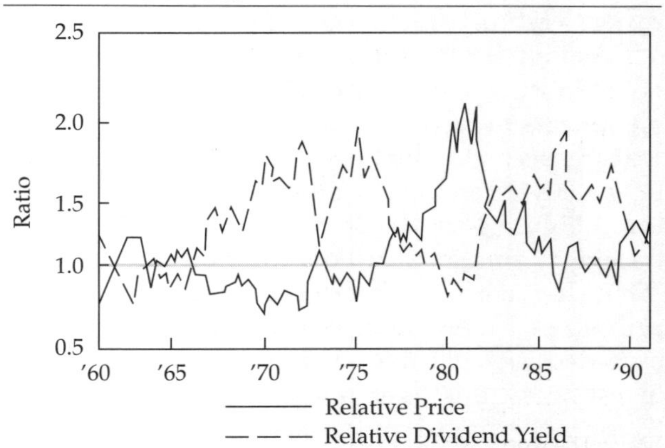

Source: Anthony E. Spare, *Relative Dividend Yield* (New York: John Wiley & Sons, Copyright 1992). Reprinted by permission of John Wiley & Sons, Inc.

[1] Anthony E. Spare (New York: John Wylie & Sons, 1992).

spond with their own investment goals and portfolio criteria. To make this assessment, analysts must communicate with company managers.

To recommend an investment, analysts must also have confidence in management, its goals, and its ability to achieve them. Analysts should also verify whether management is openly communicating with others. The worst case for an analyst is to have all the information while the rest of the market knows nothing about the stock and does not care. This situation does not make for increasing share prices.

Long-term commitment to investing. Time is an amazing healer. Wall Street focuses on quarterly reports. Most Wall Street analysts zero in on quarterly numbers and quarterly earnings surprises. Based on historical data, however, the longer the investment horizon, the greater the profitability.

Consider, for example, the experience of the large oil stocks. Investors who bought Amoco, Arco, Exxon, Phillips, Texaco, and Unocal at the high price of 1972 and held them until 1992 would have outperformed the market, as shown in Table 2. During that time, the market was up 3.5 times on price alone, but five of the six major oil stocks outperformed the market.

Figure 6 compares the total returns on Amoco and the S&P 500 since 1972. The big move in Amoco stock did not come until seven years after purchase. The next big move came in 1986, about 14 years after purchase, and another move occurred in 1989–90, 18 years after purchase. This phenomenon is not a quarterly pattern!

Figure 6. Monthly Cumulative Total Returns of $1,000 Investment—Amoco versus the S&P 500

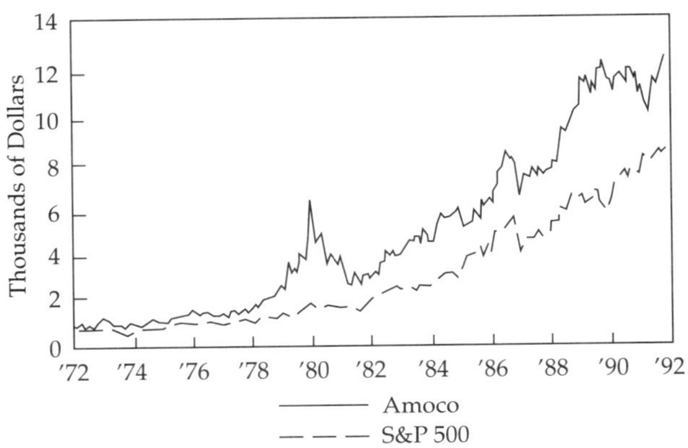

Source: State Street Research and Management Co., Inc.

Note: The worth as of October 31, 1992, of $1,000 invested in December 1972 was $12,314.35 for Amoco and $8,177.19 for the S&P 500.

Analysts and investors have experienced much trauma since 1972, but in the end, the market has rewarded those with patience and a commitment to long-term investing.

Dividend reinvestment. The oil industry pays above-market dividends; thus, reinvestment of dividends is critical to favorable long-term performance. Many oil companies, including Exxon, support dividend reinvestment by making the purchase of additional shares easy for investors.

Table 2 shows how important the value of dividend reinvestment is over time. For all six oil companies, the return on stock price is significantly lower than the total return with dividends reinvested. Amoco's stock price has increased in value by 4.5 times since 1972, for example, but its total return including dividends is 12.3 times the original price. On a total return basis, four of the six companies returned almost 50 percent more than the market. Exxon's performance was 20 times the original investment in 20 years, which is not bad performance for a nongrowth stock in a nongrowth industry. The power of dividends over time is an example of true investing. When examining valuation, analysts should keep in mind valuation on a long-term basis.

Conclusion

Oil analysts have a number of valuation methods at their disposal. Analysts add value through careful selection of appropriate valuation techniques, conscientious use of statistics, and discriminating evaluation of management. For all investors, dividend reinvestment and long-term commitment to investing are proven ways to add value to a portfolio over time.

Table 2. Return and Total Return Including Dividend Reinvestment—Selected Oil Companies

Company	1972 High Price	1992 Average Price	Return (times)	Return Including Dividends (times)
Amoco Corp.	$ 11	$ 49	4.5	12.3
Arco	20	110	5.6	14.5
Exxon Corp.	11	61	5.5	20.0
Phillips Petroleum	7	25	3.7	8.2
Texaco	39	62	1.6	7.4
Unocal Corp.	5	25	4.9	10.2
S&P 500	120	415	3.5	8.2

Source: Compustat data through Factset.

Question and Answer Session

Thomas P. Moore, Jr., CFA

Question: How do you value the differences in capital structure among firms? Is high leverage good or bad?

Moore: The effect of leverage depends on the time and place in the business cycle. When the prime rate was 14 percent in 1980 and the cost of capital was extremely high, high leverage did not benefit investors much. When the cost of capital is low, leverage is good. Many oil companies have been stretching out the maturities of their long-term debt. Some companies have maturities of 30 and 40 years on their long-term debt. As an equity investor, I use debt over equity because I want that equity leveraged up over time.

Question: How good a measure, absolute or relative, is the standardized measure of discounted future net cash flows in reflecting the economic value of variety, location, and quality of a company's oil reserves?

Moore: Not very good. I have a difficult time with dividend discount values and discounted cash flow values, because my estimates for the next quarter are usually poor. Estimates for three and five years down the road are not reliable. I may hit the right number by chance occasionally, but when estimating those numbers for the future, getting the right measure of a company is difficult. Consequently, dividend discount models are only one small piece of the mosaic.

Question: Which of the three valuation methods is appropriate for small exploration and production (E&P) companies?

Moore: I value small exploration companies today in terms of their potential. Do they have the acreage on which major discoveries could be located? Do they have the financial wherewithal to turn those discoveries into earning power?

Question: Why are certain companies, small and mid-sized producers, expensive on the basis of price to net asset value but cheap on the basis of price to cash flow?

Moore: It depends on where they are in the earnings cycle. A continuous lag exists between the time capital is invested and the return on that capital. Many companies are caught in the in-between stage. The oil industry has long cycles. Many companies that spent money in the North Sea in the early 1980s are only now benefiting from those expenditures. They might be harvesting capital already expensed and, therefore, have very high cash flows and low net asset values.

Question: In the asset value approach, when deducting the amount of debt, do you use market value or face value if debt is selling at a discount?

Moore: I have always used market value because that is what the company must pay to eliminate its debt today. The important points are to know the difference and to use whatever is comfortable for you.

Question: You mentioned $5 a barrel as a reasonable value for proven oil reserves in the ground. What do you think a proven barrel is worth in the West Siberian Basin?

Moore: Zero, although some companies are producing some of the oil. I cannot tell you whether that oil has any value yet. For investors currently willing to take the risk, it may have great benefits.

Moderator: I agree that the value of those oil reserves is zero, but they could have a call value. At least one deal of which I am aware will involve considerable Export/Import Bank financing and financing guarantees from the European Bank for Economic Recovery and Development. The result is that the company carries virtually no risk.

Moore: Most companies in business today survived the worst 10-year period in the industry's history. They are aware of and sensitive to the risk element, and they do not sign contracts without assessing risk. When possible, they insulate themselves. Many companies hedge themselves in the futures market to avoid the price risk seen in the 1970s and early 1980s.

Question: Are you comfortable with independent third-party reserve estimates by lesser known engineering firms? How should their credibility be checked?

Moore: Many outside firms do good work. Excellent reserve estimators and geologists are available, but not everyone is sending a full-blown team out to evaluate the reserves. The key factor to review when checking credibility is the client list. I keep a list of the clients of various reserve estimators. Many small reserve geologists in Lafayette, Louisiana, and Denver, Colorado, do a good job in their own regions. I would trust the reports of those companies even more than I would those of a large firm. The key is knowing the background of the firm, who it has dealt with in the past, and its current clients.

Question: What other valuation methods should be considered with respect to small E&P companies and oil field services?

Moore: There is no easy way to value the smaller E&P companies. The per-share value of potential oil and natural gas discoveries discounted for risk is a method I use. A key ingredient in the valuation of small E&P companies is an insight into their accounting methodologies.

Question: Please elaborate on valuation of acreage. Do you assign any value to probable reserves?

Moore: I usually give a 75 percent haircut to probable reserves unless extenuating circumstances exist that would enable them to be turned into developed reserves cost-efficiently or rapidly.

The Refiners—Understanding the Basics

Douglas T. Terreson, CFA
Vice President
Putnam Companies

Once the industry stepchild, the refining and marketing segment is becoming increasingly desirable as a lucrative part of an integrated oil business because of its size and significant contribution to earnings. Primary valuation methods consider book value, cash flow, valuation through comparable sales, and refinery replacement costs.

The refining and marketing segment, like the petrochemical segment of the oil business, has historically had little appeal or profitability for the integrated oil companies. Because of its size and its significant contribution to earnings, however, this segment remains a key business. In total, refining and marketing operations typically represent 40 percent of profits for the major oil companies.

In an effort to improve the quality of their total earnings streams, public and national oil companies have diversified their business mix by integrating exploration and production, refining and marketing, and chemicals operations. Some upstream entities have combined their assets with those of downstream entities to achieve integration. For example, Saudi Arabia's upstream feedstocks are combined with Texaco's downstream processing capabilities in the Star Enterprises joint venture. In a more recent agreement between Royal Dutch/Shell and PEMEX, the national oil company of Mexico, Shell's Deer Park refinery will process PEMEX-produced petroleum. The result in both situations is improved processing efficiencies and security of supply. These moves illustrate the increasing desirability of strong, integrated refining and marketing systems.

Key Industry Dynamics

Understanding supply and demand trends is important in the analysis of the refining and marketing segment. The trends shown in **Figure 1** indicate that North American production has declined during the past six years and production in the Middle East and other areas, primarily Latin America and Africa, has increased. These trends have caused a shift in the quality of crude oil supplies, because production from Latin America and the Middle East tends to be heavier and more sour than that from North America. (The term "heavier" relates to specific gravity, which is defined as the ratio of density of a crude oil to the density of water. The terms "sour" and "sweet" relate to the sulphur content of a particular crude; sour represents higher sulphur crude oil.) This decline in quality has meaningful implications for the cost of crude as a raw material and for the cost of upgrading it to produce higher value petroleum products.

The trend in price differentials, or spreads, reflects the decline in crude oil quality. **Figure 2** shows the historical price differential between West Texas Intermediate (WTI) and San Joaquin Valley crude oil. The former is a light crude oil and the latter a heavier crude oil. The increased spread between the two since 1986 reflects increasing supplies of heavy crudes.

Another supply-related issue that periodically affects crude differentials is availability. **Figure 3** illustrates the historical spread between WTI and Arab light, both of which qualify as light crudes. At the end of 1990, real or imagined supply dislocations related to the Persian Gulf crisis caused a meaningful shift in the differential. The spread remained at low levels until this particular grade of oil returned to the market.

On the demand side, the trend has been toward lighter and cleaner petroleum products. **Table 1** presents oil product consumption by fuel type for selected regions. Since 1970, demands for middle distillates and gasoline have increased in the United States, Western Europe, and Japan, and demand for fuel oil has declined.

Because light crudes yield a higher proportion of

Figure 1. World Crude Oil Production, 1970-90

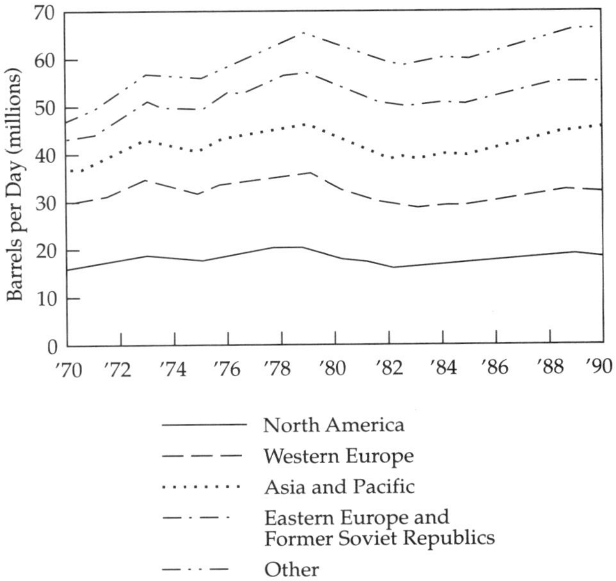

Source: Arthur Andersen & Co.

light products than do heavy crudes, the shift toward heavier crude has required a worldwide shift in the composition of refining hardware. Specifically, the distillation process, which is the first step in producing gasoline and other petroleum products, has been increasingly supplemented with sophisticated processes such as coking, catalytic cracking, and catalytic reforming. The result is a slate of petroleum products that satisfies the changing demand picture. As shown in **Table 2**, crude capacity, also known as distillation capacity, declined during the 1980s and upgrading capacity, which represents sophisticated processes, increased on both absolute and relative bases.

Economics of Refining

A simple refinery such as a hydroskimming facility is much less expensive to construct than a complex refinery. **Figure 4** and **Figure 5** illustrate typical production processes in, respectively, simple and complex refineries. A complex refinery offers the same chemical processes as a simple refinery and adds such capabilities as visbreakers, catalytic cracking units, alkylation units, isomerization units, and reforming units. (**Exhibit 1** provides a glossary of terms and definitions for the refinery schematic.)

In the 1960s, Wilbur Nelson developed the refinery complexity calculation shown in **Table 3**. The Nelson index compares the costs of building various upgrading units with the cost of building a crude distillation unit. Nelson assigned a value of 1 to crude distillation and then indexed all other processes based on relative cost to the distillation unit. For example, relative to the cost of crude distillation, the hydrocracker has a value of 10, the alkylation unit has a value of 11, and the isomerization unit has a value of 15.

Figure 2. Price Difference between WTI and San Joaquin Valley Crudes, 2Q1986–2Q'92

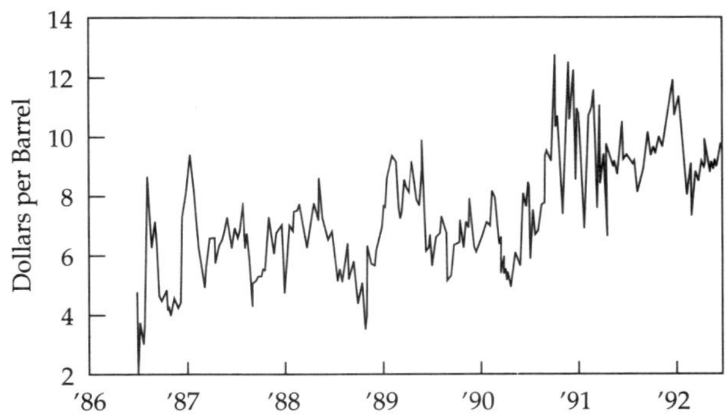

Source: Platt's *Oilgram News*.

The throughput ratio in Table 3 indicates the proportion of oil that goes through a particular process. In the simple refinery calculation, for example, the crude distillation has a throughput ratio of 1, which means that 100 percent of the crude oil fed into the refinery goes to distillation. As shown in the table, only 20 percent goes to the distillate hydrotreater. Multiplying the complexity factor by the throughput ratio yields the complexity index, which indicates the relative cost of each of a refinery's operations. In this example, the cost to build the simple refinery is approximately 25 percent of the cost to build the complex refinery. The market value of each refinery should follow accordingly.

Table 4 presents the product yield of simple, complex, and very complex refiners. Simple refiners with low complexity indexes typically have different product outputs, or yield slates, than more complex

Figure 3. Price Difference between WTI and Arab Light Crudes, 1984–2Q'92

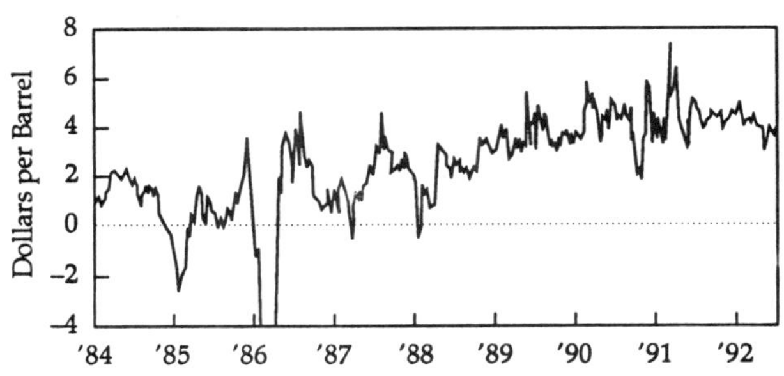

Source: Platt's *Oilgram News*.

Table 1. Oil Product Consumption by Fuel Type for Selected Regions, 1970–90
(millions of barrels per day)

Region/Fuel	1970	1975	1980	1981	1982	1983	1984	1985	1986	1987	1988	1989	1990
United States													
Gasoline	6.33	7.06	7.09	6.91	6.83	6.92	7.06	7.09	7.29	7.54	7.74	7.73	7.58
Middle distillates	3.53	3.93	4.30	4.12	3.86	3.91	4.16	4.24	4.42	4.53	4.75	4.79	4.75
Fuel oil	2.11	2.38	2.45	2.04	1.67	1.45	1.40	1.24	1.44	1.27	1.39	1.37	1.22
Others	2.73	2.95	3.21	2.99	2.93	2.94	3.11	3.13	3.08	3.32	3.30	3.34	3.32
Total	14.70	16.33	17.06	16.06	15.29	15.23	15.73	15.70	16.23	16.66	17.19	17.23	16.87
Western Europe													
Gasoline	2.43	2.95	3.26	3.12	3.14	3.20	3.23	3.24	3.39	3.46	3.58	3.62	3.65
Middle distillates	4.18	4.60	4.90	4.64	4.46	4.46	4.66	4.72	4.94	4.99	5.05	5.01	5.13
Fuel oil	4.35	4.21	3.82	3.36	2.98	2.56	2.41	2.22	2.23	2.10	2.01	2.04	2.01
Others	1.78	1.79	2.02	2.01	1.95	2.04	2.08	2.06	2.13	2.16	2.23	2.27	2.31
Total	12.74	13.55	14.00	13.14	12.54	12.26	12.39	12.24	12.70	12.71	12.88	12.94	13.10
Japan													
Gasoline	0.71	0.91	0.98	0.92	0.90	0.92	0.95	0.96	1.00	1.03	1.08	1.15	1.20
Middle distillates	0.66	1.04	1.24	1.25	1.20	1.23	1.33	1.32	1.38	1.44	1.59	1.64	1.76
Fuel oil	2.06	2.30	1.90	1.68	1.45	1.41	1.39	1.15	1.09	1.05	1.14	1.16	1.20
Others	0.57	0.76	0.81	0.84	0.85	0.83	0.95	1.01	1.03	0.98	0.99	1.05	1.10
Total	4.00	5.02	4.94	4.70	4.40	4.39	4.62	4.44	4.50	4.50	4.81	4.99	5.26

Source: Arthur Andersen & Co.

Notes: Oil consumption includes international bunkers and refinery fuel gain or loss. The category "others" comprises refinery gas, liquefied petroleum gases, solvents, petroleum coke, lubricants, bitumen, and wax.

Table 2. Refining Capacity by Type of Refining for Selected Regions, January 1980 and January 1991
(millions of barrels per day, except as noted)

	January 1980				January 1991			
		Upgrading Capacity				Upgrading Capacity		
Region	Crude Capacity	Thermal Operation	Catalytic Cracking	Catalytic Reforming	Crude Capacity	Thermal Operation	Catalytic Cracking	Catalytic Reforming
North America								
Canada	2.2	—	0.5	0.4	1.9	0.1	0.4	0.4
United States	17.7	0.4	4.7	3.7	15.6	2.0	5.4	3.9
Subtotal	19.9	0.4	5.2	4.1	17.4	2.1	5.8	4.3
Latin America	8.6	0.5	1.0	0.4	7.6	0.7	1.1	0.5
Western Europe	20.8	0.7	1.0	2.6	14.2	1.8	2.0	2.2
Eastern Europe and former USSR	13.4	NAv	NAv	NAv	14.9	NAv	NAv	NAv
Middle East	3.3	0.1	0.1	0.2	5.2	0.3	0.2	0.5
Africa	1.7	0.1	0.1	0.2	2.9	0.1	0.2	0.3
Asia and Pacific								
China	1.6	NAv	NAv	NAv	2.5	NAv	NAv	NAv
Japan	5.5	—	0.3	0.6	4.4	0.1	0.6	0.6
Other Asia and Pacific	4.9	0.2	0.3	0.4	6.0	0.5	0.4	0.6
Subtotal	12.0	0.2	0.6	1.0	12.9	0.6	1.1	1.2
Total	79.7	2.0	8.0	8.5	75.1	5.6	10.3	9.0
Upgrading capacity as a percent of crude capacity		3.1%	12.4%	13.1%		9.7%	17.9%	15.6%

Source: Arthur Andersen & Co.

Note: NAv = data not available.

Table 3. Refinery Complexity Calculation

Operation	Complexity Factor	Throughput Ratio	Complexity
Simple refinery			
Crude distillation	1.0	1.0	1.000
Gas plant	0.5	0.5	0.250
Splitter	0.3	0.3	0.090
Naptha hydrotreater	2.0	0.2	0.300
Catalytic reformer	4.0	0.2	0.600
Gasoline treater	0.5	0.2	0.075
Kerosene hydrotreater	0.5	0.2	0.075
Distillate hydrotreater	0.5	0.2	0.100
Total			2.490
Complex refinery			
Crude distillation	1.0	1.0	1.000
Gas plant	0.5	0.5	0.250
Splitter	0.3	0.3	0.090
Naptha hydrotreater	2.0	0.2	0.400
Catalytic reformer	4.0	0.3	1.200
Gasoline treater	0.5	0.4	0.200
Kerosene hydrotreater	0.5	0.3	0.125
Distillate hydrotreater	0.5	0.3	0.150
Flasher	2.0	0.8	1.560
Catalytic cracker	6.0	0.2	1.200
Hydrocracker	10.0	0.2	2.000
Alkylation	11.0	0.1	0.660
Isomerization	15.0	0.0	0.300
Total			9.135

Source: W.L. Nelson, "The Concept of Refinery Complexity," *Oil and Gas Journal* (September 1976).

refiners. The gasoline and jet fuel products from complex refineries have higher margins than the lower value products of residual fuel from simple refineries.

The relative advantage of building conversion capacity for higher margin products is not apparent with certain heavy crude oils. **Table 5** gives an example of the differences among refining margins for hydroskimming and gasoline refineries. When refining $26-a-barrel Mayan crude, the hydroskimming refinery produces approximately 10 percent high-margin gasoline, compared with 25 percent for the gasoline refinery. Because of the cut points in the distillation of this heavy crude oil, the residual fuel percentage for the hydroskimming refinery is similar to that for the gasoline refinery. Operating costs for the hydroskimming refinery are a third of those of the gasoline refinery, however, so despite the higher gasoline margin, this crude is a better fit for the hydroskimming refinery.

The difference is more readily apparent with $28-a-barrel West Texas sour crude. The volume percentage changes in the product slate are more advantageous for the gasoline refinery, despite the higher crude costs. In this case, refiners are able to reap the benefits of the decision to make the complexity or conversion upgrade.

Refinery operating margins vary according to the spread between crude oils, as shown in **Table 6.** When the spread between light cycle fuel oil, such as gasoline or distillate, and heavy cycle fuel oil, such as residual fuel oil, is $4.67, the margin is –$0.35 for hydroskimming and $1.75 for gasoline. If the product demand trends mentioned earlier continue and the spread between light and heavy goes to $8.67, then margins will improve meaningfully for the gasoline refinery but not for the hydroskimming refinery; the relative value of having constructed conversion capacity will become apparent.

The calculation of refinery operating margin can be used to value a stand-alone refinery. Oil companies' quarterly reports provide total refining and marketing earnings, and analysts divide that by the throughput. When this calculation is used, a key

Table 4. Product Yield by Refining Complexity (percent)

Product	Simple Refiners	Complex Refiners	Very Complex Refiners
Gasoline	30%	50%	65%
Jet fuel	10	19	20
Distillate fuel	20	17	25
Residual fuel	35	20	0
Fuel (gain)	5	(6)	(10)

Source: Daniel Johnston's *Oil Company Financial Analysis in Nontechnical Language*. Copyright PennWell Books, 1992.

Table 5. Components of Refining Margins

Product	Hydroskimming Refinery			Gasoline Refinery		
	Percent Volume	Unit Cost or Price	Revenue or Total Cost	Percent Volume	Unit Cost or Price	Revenue or Total Cost
Outturn West Texas sour crude ($28 per barrel)	100%	$28.00	$28.00	100%	$28.00	$28.00
Gasoline	30	32.00	9.60	50	32.00	16.00
Jet fuel	10	32.00	3.20	19	32.00	6.08
Distillate fuel	20	31.00	6.20	17	31.00	5.27
Residual fuel	35	27.00	9.45	20	27.00	5.40
Refinery fuel (gain)	5	—	—	(6)	—	—
Total outturn	100		$28.45	100		$32.75
Operating costs	100	1.00	1.00	100	3.00	3.00
Operating margins	100		(0.55)	100		1.75
Outturn Mayan crude ($26 per barrel)	100	26.00	26.00	100	26.00	26.00
Gasoline	10	32.00	3.20	25	32.00	8.00
Jet fuel	5	32.00	1.60	5	32.00	1.60
Distillate fuel	20	31.00	6.20	25	31.00	7.75
Residual fuel	60	27.00	16.20	50	27.00	12.15
Refinery fuel (gain)	(5)	—	—	(5)	—	—
Total outturn	100		$27.20	100		$29.50
Operating costs	100	1.00	1.00	100	3.00	3.00
Operating margins	100		0.20	100		0.50

Source: Johnston, *Oil Company Financial Analysis.*

Figure 4. Operations of Simple Refineries

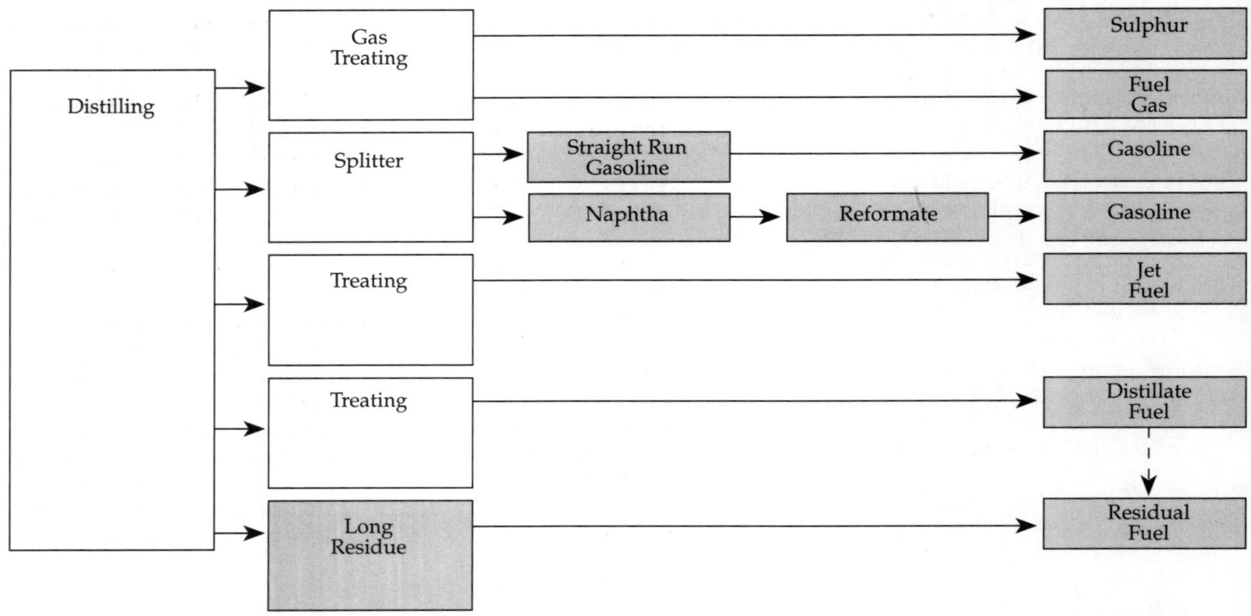

Source: William L. Leffler's *Petroleum Refining for the Non-Technical Person, Second Edition*. Copyright PennWell Books, 1985.

Figure 5. Operations of Complex Refineries

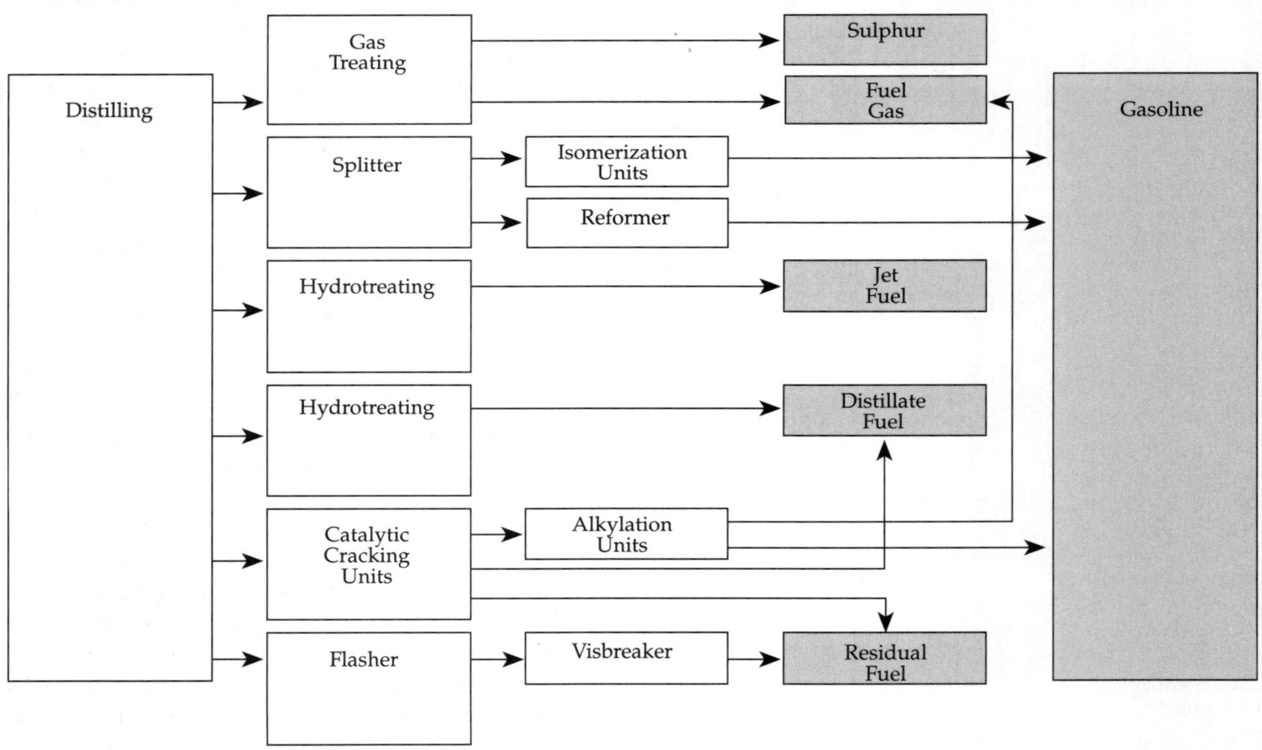

Source: Adapted from Leffler, *Petroleum Refining for the Non-Technical Person*.

Exhibit 1. Refinery Schematic Definitions

Alkylation. Combines butylenes or propylenes from the catalytic cracker with isobutane from an outside source to form alkylate, a high-octane gasoline blending component.

Catalytic cracker. Subjects heavy gas oils from the distillation tower to heat and pressure and puts the gas oils in contact with a catalyst to crack the molecular structure. Also known as *cat cracker* or *CCU.*

Distillation. Separates whole crude and reduced crude feedstocks into various products for blending, sale, or further processing.

Distillate fuel. Product from the distillation tower and the catalytic cracking unit. Also known as *diesel fuel* or *furnace oil.*

Flasher. Subjects residue from the bottom of the distillation tower to heat and reduced pressure to produce heavy and light cycle oil, which can be sent to the cat cracker.

Gas treating. Removes impurities such as sulphur from the gas stream and separates commercial gases such as ethane, propane, butane, etc.

Hydrotreater. Uses hydrogen to remove sulphur, organic nitrogen, carbon residue, and metals from feedstocks or end-products.

Isomerization. Converts normal paraffins (butane, pentane, and hexane) into iso-paraffins. The converted product is a valuable, high-octane gasoline blending component.

Long residue. See *Residual fuel.*

Reformer. Upgrades naphtha to reformate, a premium gasoline blending component. Also produces hydrogen, an important refinery reacting agent.

Residual fuel. Low-quality, heavy fuel also known as long residue. Feedstock for the flasher.

Visbreaker. Takes pitch from the flasher and thermally cracks it. The pitch can then be reflashed and combined with a diluent and marketed as residual fuel.

factor to monitor is the direction of volume, primarily for a stand-alone refinery.

Fuel switching has been a hot topic. Refiners must determine on a daily basis how to operate the refinery to optimize profits in light of what the market is paying for different products. **Figure 6** compares different petroleum products on the basis of British thermal unit (Btu) equivalency. For a given value of natural gas, refiners can determine the equivalent value of other products available to electrical and industrial users. Over time, the analysis might signal a shift among gasoline, distillate, or fuel oil, and the refiners decide whether that shift will optimize their profits.

Analysts also look at inventory reports like the one shown in **Table 7**. Every Wednesday, the *Wall Street Journal* publishes inventories for the previous week on the commodity page. From one week to the next, the report might show inventories of crude oil increasing by 10 million barrels, for example, and gasoline inventories decreasing by 5 million barrels. With these example inventory changes, crude oil prices would decrease and gasoline prices would increase, which would improve the margin for refiners. Long-term demand drives this trend, and at some point, quarterly earnings will reflect the improvement.

Table 6. Refinery Operating Margins

Spread	Hydroskimming	Gasoline
West Texas sour crude		
4.67 spread	$(0.35)	$1.75
8.67 spread	0.70	4.13
Mayan crude		
4.67 spread	0.20	0.50
8.67 spread	0.65	1.65

Source: Johnston, *Oil Company Financial Analysis.*

The second item of interest on the inventory report is gasoline inventories, consumption, and supply by Petroleum Administration for Defense District (PADD) region (Table 7). If West Coast refining and marketing margins are improving because of long-term demand trends and/or short-term capacity constraints, analysts and portfolio managers can make several investment decisions. For instance, Chevron, Unocal, and Arco offer ways to profit from improvements in West Coast refining and marketing profitability. If Midwest inventories are drawn down or consumption improves, analysts might recommend USX-Marathon or Amoco.

Analysts look closely at apparent gasoline consumption (middle panel of Table 7). Because consumption is so volatile, the data are reported as four-week moving averages. Dividing inventories by consumption yields gasoline days of supply (third panel of Table 7), which analysts compare with last year's levels.

The majority of U.S. oil companies use LIFO to calculate inventory. LIFO approximates replacement costs for most of the major integrated oil companies in North America today. It understates the value of inventories during rising prices, however, and overstates the value during falling prices. The market value of inventories in excess of book value is usually disclosed in the financial footnotes as a LIFO reserve. When large differences exist between

Figure 6. Equivalent Fuel Values for Selected Hydrocarbons

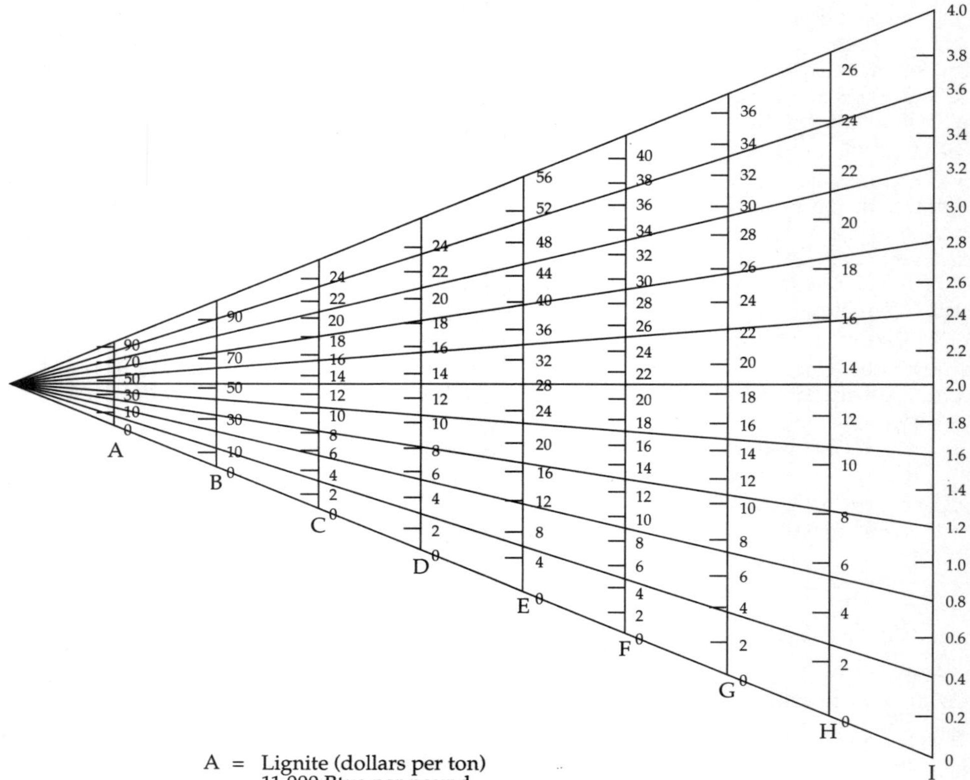

A = Lignite (dollars per ton)
11,000 Btus per pound

B = Subbituminous Coal (dollars per ton)
12,600 Btus per pound

C = Number 6 Fuel Oil, 0.3% Sulfur (dollars per barrel)
151,500 Btus per gallon

D = Number 6 Fuel Oil, 2.5% Sulfur (dollars per barrel)
153,000 Btus per gallon

E = Number 2 Fuel Oil (cents per gallon)
140,000 Btus per gallon

F = Butane (cents per gallon)
103,300 Btus per gallon

G = Propane (cents per gallon)
91,600 Btus per gallon

H = Ethane (cents per gallon)
66,000 Btus per gallon

I = Natural Gas (dollars per million Btus)

Source: Leffler, *Petroleum Refining for the Non-Technical Person.*

LIFO value and market value, companies can create inventory profits by liquidating their excess inventories.

Many foreign oil companies use the FIFO inventory method. FIFO produces relatively higher earnings as prices rise and lower earnings as prices fall. It effectively magnifies the results of business cycles in reported earnings.

Of the two methods, LIFO accounting generates the more realistic earnings, because current cost for inventories is used in the calculation, but FIFO provides a more realistic value of inventories. In times of stable prices, both systems ultimately yield the same earnings and inventory values.

Some companies use the average-cost inventory method or a combination of LIFO, FIFO, and average cost to smooth out inventory costs, and some companies use the futures market to smooth inventory costs.

After the immediate and sharp increase in crude oil prices related to the Persian Gulf crisis in 1990, Shell distributed an article explaining why, although gasoline prices increased almost immediately when crude prices moved up, oil companies did not necessarily

Table 7. First Boston Corporation Energy Group—Flash Analysis, United States (American Petroleum Institute statistics)

	10/23/92	10/16/92	Change	10/25/91
Inventories (thousands of barrels)				
Total				
Crude oil	330,354	327,260	3,094	347,780
Gasoline	208,973	209,756	(783)	205,694
Distillate fuel	133,276	131,516	1,760	134,802
Residual fuel	43,726	45,069	(1,343)	45,590
Utilization rate (percent)	89.2	88.3	0.9	81.5
Gasoline by region				
PADD 1 (East Coast)	57,585	56,411	1,174	55,127
PADD 2 (Midwest)	58,995	60,529	(1,534)	54,746
PADD 3 (Gulf Coast)	57,153	57,819	(666)	60,541
PADD 4 (Rocky Mountains)	5,767	5,726	41	5,995
PADD 5 (West Coast)	29,473	29,271	202	29,285
Total gasoline	208,973	209,756	(783)	205,694

	Four-Week Average			Year-to-Date Average		
	10/23/92	10/25/91	Percent Change	1992	1991	Percent Change
Apparent Gasoline Consumption[a] (thousands of barrels)						
PADD 1	1,047	1,002	4.5%	991	976	1.5%
PADD 2	1,718	1,880	–8.6	1,748	1,726	1.3
PADD 3	3,229	3,023	6.8	3,136	3,063	2.4
PADD 4	221	232	–4.7	231	252	–8.3
PADD 5	1,164	1,295	–10.1	1,230	1,279	–3.8
Total	7,379	7,432	–0.7	7,336	7,296	0.5

	10/23/92	10/25/91
Gasoline Days Supply		
PADD 1	55	55
PADD 2	34	29
PADD 3	18	20
PADD 4	26	26
PADD 5	25	23
Total	28	28

Source: First Boston Corporation.

[a]Calculated as follows: Last week's inventory + Production + Imports – Current week's inventory.

show immediate gains.[1] For the most part, third-quarter 1990 earnings in refining and marketing were washed out because feedstock costs increased faster than product costs. In the fourth quarter, companies increased product costs slightly, which improved margins. The converse occurred in the first quarter of 1991, when crude prices decreased sharply. Oil companies made healthy profits because product prices decreased less quickly than crude prices. Over time, margins even out, but the volatility during that period led to swings in short-term profits.

[1]Shell International Petroleum Co., Ltd., "High Oil Prices and Stock Gains—The Illusion of Profit" (London, October 1990).

Regulatory, Environmental, and Geographical Issues

The legislative environment and the effect of environmental regulations are the biggest issues facing the refining and marketing segment, and the oil business in general. These issues hold particular significance for the overseas market, in which the industry is considering more capacity additions. Two key issues in the United States are the 40-miles-a-gallon average efficiency standard for automobiles and the possibility of a tax on gasoline. If these measures go into effect, they could lead to a decline in gasoline demand similar to that from 1980 to 1985.

The estimated cost of environmental regulations ranges between $10 billion and $20 billion for the industry. The large integrated oil companies estimate that 20–40 percent of capital expenditures for downstream refining and marketing will be related to environmental regulations. Oxygenate-related capacity will account for the majority of these expenditures, and the outlay should yield some return on capital, but if 30 percent of downstream capital expenditures have a zero return on capital, earning a total return in excess of the cost of capital will be difficult.

Although some analysts believe oil companies have an interest in presenting high estimates of environment-related costs, no one really knows the magnitude or the eventual effect of the spending. Present and proposed environmental regulations may make low-conversion plants prohibitively expensive to operate. If U.S. environmental regulations, including the Clean Air Act, become more strict toward refiners, the one million barrels currently at risk of closing will multiply during the next few years.

In November 1992, 39 U.S. cities were required to use gasoline with 2.7 percent oxygen by weight, which could be accomplished by blending 15 percent (by volume) methyl tertiary butyl ether (MTBE) or 11 percent (by volume) ethanol. For the most part, this change has been absorbed into the gasoline market. Although there is a shortage of MTBE capacity, refiners produced the commodity during the spring and summer of 1992 and stored it. They were able to meet daily demand during the winter through a combination of production and drawdowns from storage. MTBE prices are now about 80–85 cents a gallon. Construction of new MTBE capacity during the next two to three years should add another 200,000 barrels a day to meet additional demand in 1995 and 1996.

Geographical considerations also affect the refining and marketing business. Oil companies benefit from the lower costs associated with being located at or near growing and profitable markets. In North America, growth in demand for petroleum products has been anemic, and profitability has suffered. Shell, however, which originated in the Far East in the early 1900s, has access to those high-growth markets today. In addition, CALTEX, a joint venture between Chevron and Texaco, is a very attractive downstream entity, because it also has access to those Far East markets.

The economic availability of oil and natural gas is a key consideration. Low transportation costs are an economic advantage. Transportation via pipeline connections and/or water access is critical. The speed and cost-effectiveness of modern distribution systems and intercompany supply agreements can offset transportation and feedstock costs at the point of origin. Utilities (including electricity), plant fuel, and a reliable water supply are also important to refiners. Many petrochemical plants in the Gulf Coast region benefit from access to natural gas. They have built electricity-generating plants, which allows them to generate all the electricity they need and sell any excess to power companies.

Ensuring a supply of skilled labor is important. It is more of a problem overseas than in the United States, where the large concentration of refining capacity in the Gulf Coast area alleviates the problem. In some overseas areas, labor agreements present problems when a company is building new capacity or adding to existing capacity.

Valuation

Several approaches are used to value oil companies. Some of these methods are illustrated in **Figure 7**, which shows five such measures relative to the S&P 500 for 1987 to late 1992 for several stand-alone refineries. By most of these measures, relative total return has been poor since 1989. The companies have been burdened by the investment requirements of the Clean Air Act and by weak demand.

- *Book value.* The book value of oil and natural gas found on a balance sheet bears little relationship to asset value. Typically, the book value of the assets in the oil business, upstream or downstream, differs drastically from replacement cost. Book value of refining and storage facilities, however, represents actual construction costs or purchase price less depreciation, depletion, and amortization.

  Although refineries typically sell at a discount to book value, their multiples of book value may range from 0.75 to 1.25 times based on the complexity and profitability of the refinery, as well as the industry and market environment.

- *Cash flow.* The ratio of price to cash flow is probably the most consistent and reliable valuation measure for this industry. Many analysts use ratios of absolute price to cash flow and of relative price to cash flow as the valuation methodology of choice. Integrated oil companies typically sell between 5.0 and 6.0 times cash flow, and refiners sell between 3.5 to 5.0 times. They sell at a discount because the

Figure 7. Valuation Methods for Refining Companies
(relative to the S&P 500)

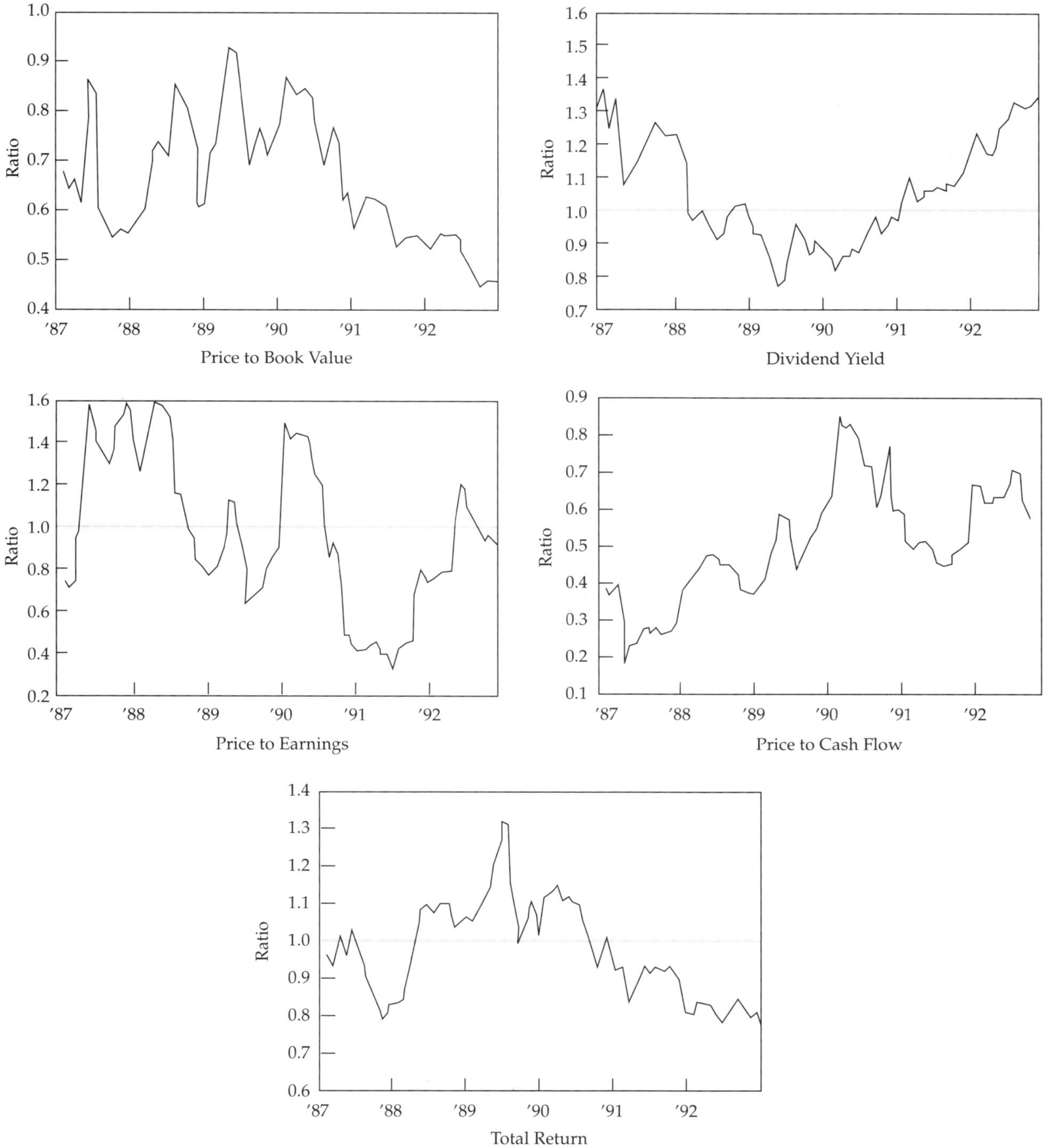

Source: Smith Barney.

quality of earnings for refining companies is low compared with that of integrated oil companies, which have upstream capabilities that can offset downstream losses in periods of poor profitability. Because of the capital-intensive nature of the refining and marketing business, operating profit multiples for these companies include a large component of depreciation. Cash flow provides a better approximation of value because it is much more stable than operating profit from period to period.

Valuation through comparable sales. Analysts determine comparable sales by computing dollars paid per daily barrel of distillation capacity and adjusting that figure for complexity. For comparison purposes, they can then use the complexity index in **Figure 8** to adjust for differences in hardware among the companies. The measure used, dollars per barrel per stream day, is based on 330 days rather than the 365 days in a normal year.

Comparable sales of refineries can be difficult to find, because the market is not active. During the past several years, few refining and marketing assets have been sold. Each year, five to seven refineries change ownership, and a similar number close down. More closings will occur in the future. Some analysts estimate 800,000 to 1,000,000 barrels a day of refining capacity, the equivalent of 300,000 barrels a day of gasoline capacity, will shut down during the next four or five years.

Refinery replacement cost. Building a modern refinery today in the United States costs between $5,000 and $6,000 per daily barrel, a prohibitive cost considering the lead time necessary for engineering and construction of the facility. For many companies, buying refining capacity is cheaper than building it. The prices paid in transactions in the late 1980s represented only 60–70 percent of replacement costs.

Figure 8. Refinery Value
(dollars per barrel per stream day)

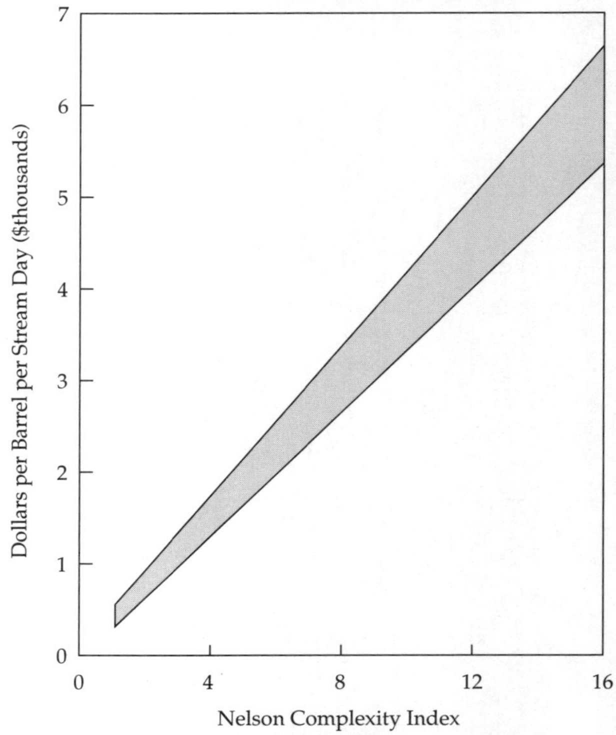

Source: Johnston, *Oil Company Financial Analysis.*

Question and Answer Session

Douglas T. Terreson, CFA

Question: Is the calculation of the theoretical crack spreads useful in predicting companies' operating profits?

Terreson: Yes, but that calculation is geared toward a single refining system. For example, the calculation used today would be appropriate for Valero Energy, which derives the majority of its earnings from refining and marketing. The situation changes when one is analyzing integrated oil companies and earnings forecasts for worldwide operations with varying degrees of complexity. Modeling refining systems accurately on a geographical basis is difficult for companies such as Royal Dutch/Shell, which has about three million barrels of refined product capacity on a worldwide basis. Dividing output by operating earnings is the best approach.

Question: Do you think more refiners will move overseas to beat Environmental Protection Agency regulations and use old refineries as storage or reformulation centers?

Terreson: Refining capacity will continue to move overseas. Today, the United States imports about 120,000 barrels a day of gasoline, but as environmental regulations become more onerous, that number will increase. Last year's study by Energy Security Analysis indicated that other major consumers of petroleum products worldwide are not likely to adopt the same oxygenate standards as the United States. If that is the case, U.S. operating costs will rise relative to those overseas. As a result, the United States will produce less and import more petroleum products.

Question: Amerada Hess and Repsol are reconfiguring their refineries to produce lighter products. Cash flow for each company is projected to rise substantially in 1994 and 1995. Historically, have investors made money by purchasing stocks on the basis of future refinery upgrades?

Terreson: You must examine both companies and their fundamentals in the context of the overall industry environment. If you believe refining and marketing will be a difficult business next year, then margins will be relatively flat based on capacity increases. You must assign the same profitability measures to both companies. On the other hand, Amerada Hess is taking 40,000 barrels of residual fuel and upgrading it to gasoline. So, although margins might be flat, these companies' profitability must increase, because gasoline margins are much better than residual margins.

Question: What impact will the return of Iraqi oil have on U.S. refiners, especially complex refiners such as Valero?

Terreson: The Iraqi oil removed from the market was heavier than the incremental barrel that has since replaced it. If Valero has been purchasing a heavier feedstock than other companies, and if Iraqi oil becomes available, then the price spread will have to decrease. For example, if the Iraqis add 1.5 million barrels a day of capacity, the resultant product slate from the incremental barrel will be heavier than it now is. If Valero is buying a similar grade of crude, its feedstock costs will decrease.

Question: Given the addition of oxygenates to the gasoline pool, and assuming no growth in gasoline demand, how much refining capacity needs to close to maintain or improve the capacity utilization rate?

Terreson: A recent poll indicated almost a 500,000-barrels-a-day increase in gasoline capacity between 1992 and 1994. Oxygenate capacity will be about 130,000 of the 500,000 barrels. Gasoline demand in the United States today is about 7.3 million barrels a day, so the new capacity amounts to a 6 percent increase in gasoline capacity, including oxygenate. U.S. gasoline demand has not grown by 6 percent in the past 10 years, and it is not expected to grow by 6 percent in the next 5 years. If demand does not grow, utilization rates will decline with the addition of oxygenates, including MTBE, to the gasoline supply.

Question: What impact will Middle Eastern refineries have on the U.S. refining industry during the next five years?

Terreson: Europe consumes much of the output of Middle Eastern refineries; the United States imports only about 100,000 barrels a day from the Middle East. The national oil companies in the Middle East have sought to integrate their upstream and downstream systems. Texaco's Star Enterprises joint venture is one example. Middle Eastern refineries have an aggressive expansion program under way. As long as they meet the product specifications required in the United States, they can compete in the market.

Understanding Petrochemical Operations

James F. Clark
Vice President and Equity Research Analyst
First Boston Corporation

Although the petrochemical industry contributes important revenues and earnings to the oil industry, analysts must use caution in making long-term valuations. This segment of the industry is cyclical and tends to attract capital at the wrong times.

The petrochemical segment of the oil business is generally not well covered by Wall Street's integrated-oil analysts. For most of us, the petrochemical business is a small, relatively unknown segment best covered by tracking current economic trends. This analytical oversight, however, has proved costly during the past few years, when segment profitability was declining dramatically. Analysts should take the time, therefore, to assess the value of this sector as a distinct business.

Why Chemicals Matter

The petrochemical sector deserves stand-alone analysis for four reasons. First, almost all integrated oil companies are engaged in the petrochemical business. **Table 1** shows 14 integrated oil companies and the petrochemicals they produce. Ethylene, polyethylene, styrene, polyvinyl chloride, and other hydrocarbon derivatives represent the bulk of production in this industry. These products are only a portion of the petrochemical market, however. Although not covered in this presentation, aromatics, aliphatics, and alcohols are also classified as petrochemicals.

Second, many oil companies derive a significant portion of earnings from petrochemical operations. **Table 2** shows the degrees to which chemical earnings contribute to total operating earnings for some major oil companies. For the group, petrochemicals during the 1986–91 period averaged 18.3 percent of earnings—a significant amount. The spectrum of exposure is wide: At one extreme, two companies, Amerada Hess and USX-Marathon, have no exposure; at the other extreme, Occidental Petroleum is one of the biggest polyethylene and polyvinyl chloride producers in the country, with an average of 45.3 percent of its earnings from petrochemicals during the 1986–91 period.

Third, the volatility of petrochemical contributions has been dramatic. During the peak years between 1987 and 1989, petrochemicals contributed approximately 20–30 percent of oil company earnings. In 1992, a trough year, that figure was 5–10 percent.

Finally, the petrochemical industry is poised for a recovery. Almost all chemicals are in excess supply, so pending economic recoveries should help utilization rates significantly.

Monitoring Margins

Tracking margins is important in understanding the industry's supply-and-demand fundamentals, because knowing margins helps establish realistic estimates for the following year.

Ethylene feedstock margins have exhibited considerable cyclicality, as **Figure 1** illustrates. The peak margin year was 1988. In that year, with a fairly robust U.S. economy, both capacity to produce and demand for ethylene in the United States was 36 billion pounds—a capacity utilization rate of 100 percent. Ethylene margins for the year were 23.5 cents a pound. In 1992, capacity was 47 billion pounds and the average utilization rate was only 83 percent. This decline has had a deleterious effect on margins, which now average about 9 cents a pound. With a 31 percent addition to volume and only a 13 percent addition to demand, the petrochemical sector is in a classic cyclical overcapacity situation.

The ethylene market should improve from current levels during the next three to five years. Between 1992 and 1996, gross capacity for ethylene should increase 5 percent. Closings, such as Phillips Petroleum's 1992 closing of a facility producing 400

Table 1. Chemical Products by Producer

Company	Ethylene	High-Density Polyethylene	Vinyl Chloride	Polyvinyl Chloride	Propylene	Polypropylene	Styrene	Polystyrene	MTBE[a]
Domestic integrated oil companies									
Amerada Hess									×
Arco Chemical							×	×	×
Kerr-McGee									×
Occidental Petroleum	×	×	×	×	×				
Phillips Petroleum	×	×			×	×			×
Unocal Corp.									
USX-Marathon Group									×
International integrated oil companies									
Amoco Corp.	×				×	×	×	×	×
British Petroleum	×								
Chevron Corp.	×	×			×		×	×	×
Exxon Corp.	×	×			×	×			×
Mobil Corp.	×	×						×	×
Royal Dutch/Shell	×				×	×			
Texaco	×				×				×
Independent refiners									
Ashland Oil									×
Diamond Shamrock									×
Sun Co.	×								×
Valero Energy									×

Source: Chemical Marketing Reporter.

[a]Methyl tertiary butyl ether.

Table 2. Chemical Earnings as a Percent of Total Earnings, 1986–91

Company	1986	1987	1988	1989	1990	1991	Average 1986–91
Domestic integrated oil companies							
Amerada Hess	0.0%	0.0%	0.0%	0.0%	0.0%	0.0%	0.0%
Atlantic Richfield	15.3	27.8	41.7	26.4	18.6	22.2	25.3
Kerr-McGee	45.4	36.4	45.4	42.8	33.8	37.8	40.3
Occidental Petroleum	11.8	26.6	68.6	76.1	51.7	36.8	45.3
Phillips Petroleum	29.8	56.5	51.7	41.9	21.3	23.9	37.5
Unocal Corp.	11.9	6.6	7.5	6.7	5.3	7.2	7.5
USX-Marathon Group	0.0	0.0	0.0	0.0	0.0	0.0	0.0
International integrated oil companies							
Amoco Corp.	32.0	29.6	30.9	28.6	7.6	4.1	22.1
British Petroleum	8.9	8.0	17.1	18.5	4.4	NM	11.4
Chevron Corp.	7.3	17.0	25.5	21.7	0.9	7.3	13.3
Exxon Corp.	8.3	14.3	21.8	19.6	8.5	8.0	13.4
Mobil Corp.	5.7	14.3	24.4	23.7	7.6	5.2	13.5
Royal Dutch/Shell	16.9	24.4	30.5	25.6	14.0	2.0	18.9
Texaco	3.0	6.3	11.7	19.3	2.4	1.3	7.3
14-company average	14.0	19.1	26.9	25.1	12.6	12.0	18.3

Source: Company annual reports.

Note: NM = not a meaningful figure.

million pounds a year, will result in a lower net figure. At the same time, demand should grow approximately 11 percent. Margins for ethylene, which increased to approximately 12 cents a pound in November 1992, are beginning to show indications of a possible improvement.

The high-density polyethylene (HDPE) market has weakened considerably during the past few years, as shown in **Figure 2**. At the peak of the market in 1988, domestic capacity reached 8.4 billion pounds a year, and capacity utilization was at 96 percent. Margins on HDPE reached 26 cents a pound and have been in an erratic fall since then.

Integrated producers, benefiting from the capital attracted by increased margins on ethylene and polyethylene, created an even more dramatic increase in capacity, 32 percent between 1987 and 1992. Demand increased only 21 percent, however, so margins declined to 8 cents a pound. Capacity utilization rates in the industry are now 85 percent.

The current outlook in the polyethylene industry is poor. Capacity should increase 13 percent by 1996, whereas demand is forecasted to grow only 12 percent for the same period, which means the industry will most likely maintain the mid-1980s' operating rate. In the absence of increased economic activity, polyethylene margins will remain low. Because of the poor outlook for the market, Occidental plans to

Figure 1. Ethylene Feedstock Margins

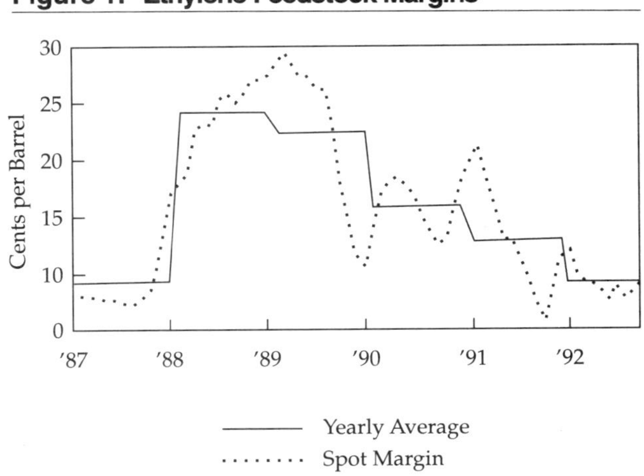

Source: Domestic Petrochemical Monthly, First Boston Corp.

Figure 2. High-Density Polyethylene Feedstock Margins

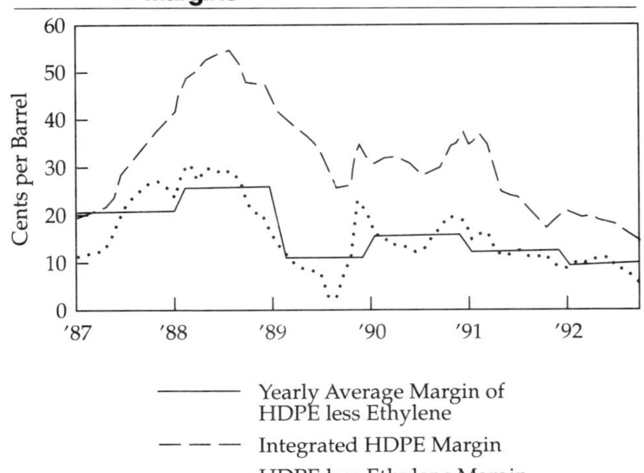

Source: Domestic Petrochemical Monthly, First Boston Corp.

close one of its small, inefficient plants in the Southeast. If margins continue to weaken—and they have appeared weaker since September 1992—more closings may follow those already scheduled.

The import situation is indicative of the overcapacity in the ethylene/polyethylene market. The United States is normally a net exporter of ethylene and its derivatives. In 1988, U.S. net exports of ethylene products—including glycols, dioxides, and oxides—were 700 million pounds. Through August 1992, the country was exporting at the rate of 4.2 billion pounds a year, a sixfold increase over 1988's net exports. The increase is the result of current U.S. economic conditions, new capacity coming on-line, and the formerly weak dollar.

Polyvinyl chloride (PVC) has experienced a relatively stable margin, as illustrated in **Figure 3**, in part because it has not attracted capital investment and, accordingly, has avoided overcapacity. Utilization rates have remained near 90 percent. Some additional plants may come on-line, but managements, burned by the petrochemical cycle in the past, have not planned many new plants. Almost all of the erosion in PVC margins is attributable to a weaker-than-expected economy.

Capital allocated to chemicals as a percentage of total capital spending at the integrated oil companies is currently at a 12-year low. The reasons are that (1) a great deal of money is being allocated to upstream activities, and (2) oil companies are allocating, on average, 24–26 percent of capital to refining and marketing, much of which is for compliance with the Clean Air Act. As a result, expansion of capacity in chemicals is being curtailed, which is probably a positive sign for margins two or three years in the future.

Forecasting Earnings

Analysts can use several approaches to forecast petrochemical earnings. One is to develop sensitivity tables to show how earnings will be affected by a 1-cent-a-pound move. Using this approach, analysts can estimate how much earnings will change for a given production estimate. Although quick and helpful, this technique is not really the best for an oil analyst, because it omits other key variables.

An alternative approach is to forecast earnings based on gross margin and volume estimates. **Table 3** illustrates this approach for Phillips. The technique uses margins, volume, depreciation, and operating costs to determine operating income for a given segment. The table may appear simple, but accumulating the data, which are not always provided by the companies, makes constructing the table more difficult than it appears. In this industry, depreciation figures, however, are well known because companies disclose their per-unit depreciation. Table 3 also shows depreciation and operating costs on a per-pound basis. In Phillips's case, operating costs decreased dramatically when it added capacity that was more efficient than previous facilities and thus reduced labor costs.

This example covers only the ethylene/polyethylene portion of the business, but the same analysis should be applied to propylene and other products. Analysts can usually estimate the total petrochemical segment's earnings within 15–20 percent by analyzing each product in this way. The model has a tendency to overstate earnings, but it provides a realistic value for the petrochemical segment of an integrated oil company, and it is better than ignoring the entire segment, basing a plug figure on current economic trends, or hoping the segment performs as the investor relations staff predicted.

Valuation

It is futile to value companies such as Occidental Petroleum, which derives 45 percent of its earnings on a normalized basis from petrochemicals, solely as an integrated oil company. An alternative approach is to use a segment earnings model—that is, to treat the component operating companies or divisions as separate businesses. Market multiples are estimated for each segment based on its industry multiple. The market multiples for each business segment are then combined on a weighted-average basis to determine the overall company multiple, from which a target stock price can be determined.

Table 4 illustrates segment valuation applied to

Figure 3. Polyvinyl Chloride Feedstock Margins

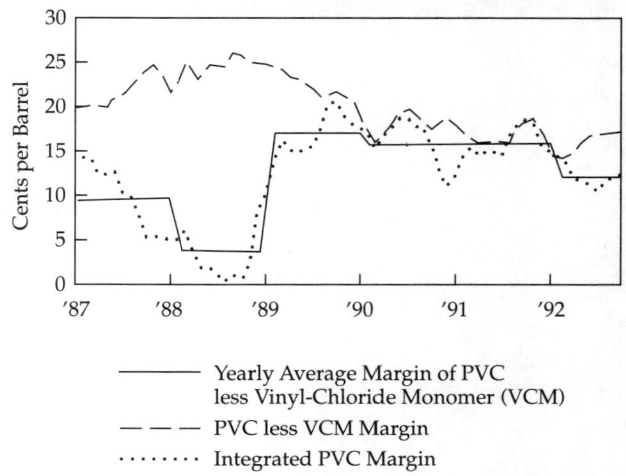

Source: Domestic Petrochemical Monthly, First Boston Corp.

Table 3. Calculating and Forecasting Petrochemical Earnings, Phillips Petroleum
(dollars, except as noted)

Item	1990	1991	1992	1993[a]	1994[a]
Ethylene volume (millions of pounds)	1,500	1,500	1,500	1,500	2,000
Polyethylene volume (millions of pounds)	1,500	1,500	1,500	1,500	2,000
Polyethylene price/pound	0.39	0.32	0.26		
Ethylene price/pound	0.24	0.21	0.17		
Feedstock margin/pound	0.15	0.11	0.08	***0.09***	***0.10***
Gross olefin margin (millions)	220.00	167.00	122.00	135.00	200.00
Depreciation/pound	0.01	0.01	0.02	***0.02***	***0.02***
Depreciation (millions)	20.00	20.00	25.00	25.00	25.00
Operating costs/pound	0.07	0.05	0.04	***0.05***	***0.06***
Total operating costs (millions)	100.00	72.00	67.00	75.00	120.00
Operating income (millions)	100.00	75.00	30.00	35.00	55.00
Taxes (millions)	(35.00)	(26.00)	(11.00)	(12.00)	(19.00)
Net income (millions)	65.00	49.00	20.00	23.00	36.00

Source: First Boston Corp. estimates.

Note: Key assumptions are in bold italic.

[a]Estimate.

Occidental. As shown, Occidental has four segments: exploration and production, natural gas transmission, chemicals, and coal. The aggregate of the segments' multiples gives Occidental an overall multiple for 1993 of 15.6. If the analyst believes Occidental's earnings will increase next year (to $1.30, for example, which is what some analysts are predicting), then the model indicates that the stock, currently trading at $17, is undervalued.

Basing earnings estimates for petrochemical companies such as Occidental on recent history can be misleading because of the cyclical nature of commodity chemicals. Thus, the 1994 figures in Table 4 assume a normal PVC cycle in an economic recovery, which would add about 2 cents a pound per month to the overall price of PVC, for an average annual increase of 12 cents a pound. An increase of this size would boost the company's earnings from the chemical segment considerably (from $0.70 a share to $1.45 a share) and have a leveraging effect, as the model shows, on total earnings per share. Analysts who think chemicals are not worth monitoring because they will never turn around, and who thus avoid Occidental, could miss a well-performing stock.

Conclusion

The petrochemical industry is an important segment of the oil industry because it contributes significant revenues and earnings. Tracking the segment is important, but analysts must be careful about long-term valuation, because the petrochemical sector is cyclical and tends to attract capital at the wrong times.

Table 4. Segment Earnings Model, Occidental Petroleum
(millions of dollars, except as noted)

Item	Current Industry Group Multiple	1993 Earnings	1993 Earnings (percent)	1993 Contribution Relative Multiple	1994 Earnings	1994 Earnings (percent)	1994 Contribution Relative Multiple
Reported net income							
Exploration and production	20.0	$230.00	26.7%	5.3	$310.00	27.0%	5.4
Natural gas transmission	16.3	300.00	34.9	5.7	310.00	27.0	4.4
Chemicals	12.0	310.00	36.0	4.3	510.00	44.3	5.3
Coal	10.0	20.00	2.3	0.2	20.00	1.7	0.2
Income before corporate charge		860.00		Aggregate 15.6	1,150.00		Aggregate 15.3
Corporate charge		(80.00)			(80.00)		
Total		$335.00			$610.00		
Average shares outstanding (millions)		299			299		
Reported net income/share (dollars)		$ 1.12			$ 2.04		
Net operating income/share (dollars)		1.12			2.04		
Target stock price (dollars)		17.47			31.18		

Sources: Company reports; First Boston Corp. estimates.

Question and Answer Session

James F. Clark

Question: Besides growth in gross domestic product, can any other macro or micro factors help forecast demand in the chemicals sector?

Clark: GDP is the principal factor we use to predict overall demand growth; other factors are useful in predicting growth for the individual chemicals. With PVC, for example, the number of housing starts is a good indicator, probably a more useful tool than GDP. World GDP is a better indicator for polyethylene than domestic GDP, because this chemical has a worldwide market. Long-term weather forecasts can be used for such products as ethylene glycol, which goes into antifreeze. Growth for methyl tertiary butyl ether (MTBE), which is added to gasoline to comply with the Clean Air Act, is currently unrelated to the economy. It has a government-mandated market that will grow by a factor of three to four times between now and 1996. Whether MTBE will avoid the commodity chemical cycle is not clear. So much capacity building is in progress that producers will probably not realize enormous profits. If the Clean Air Act Amendments of 1990 had not been enacted, MTBE growth would be tied only to gasoline consumption and GDP would be the primary indicator. Over the long term, however, these commodities are all related to economic factors.

Question: Please address the recent trade-off between use of liquid natural gas and use of naphtha. What impact did this have on gas markets?

Clark: The recent increase in prices of gas liquids favored a switch to naphtha, where possible, by U.S. ethylene crackers. Switching capability is limited, however, because of the inflexibility in the ethylene system. It is based primarily on a mix of gas liquids—ethane and propane. If gas prices were to increase considerably, more producers would switch from gas to naphtha, because naphtha prices have been holding relatively stable. Ethylene crackers with the capability to do so, some of the more flexible producers, use naphtha now, because on an input basis, they receive about a half-cent-a-pound advantage.

Question: Is overcapacity in the Far East a concern?

Clark: The Far East continues to build capacity, but based on projections from Royal Dutch/Shell, most of the capacity being built should be absorbed in that region. In fact, the Far East may need even more capacity than that currently planned. The capacity building would be a concern only if the growth rate in the Pacific economies, particularly in the Pacific countries of the Organization for Economic Cooperation and Development and the less-developed countries, suddenly failed to continue at 5 percent a year.

Question: How is recycling affecting these businesses, particularly high-density polyethylene, low-density polyethylene, and polypropylene?

Clark: The polystyrene market is the most negatively affected by recycling. For the other markets, recycling is not a particularly large factor. Most recycling is diverted into uses other than the original uses. For example, the high-density polyethylene in plastic bottles will end up in a low-density polyethylene use after recycling.

Question: What is the industry ratio of domestic to foreign sales?

Clark: Overall, for the six major international oil companies, the ratio of U.S. to foreign sales averages 50/50. For Exxon and Royal Dutch/Shell, the bulk of revenues comes from outside the United States; for Chevron and Texaco, the bulk of revenues comes from within the United States. In domestic integrated oil companies, probably 70 percent of segment earnings are from domestic operations. One exception is Arco, which manufactures propylene oxide and MTBE in France.

Question: What are the prospects for foreign sales growth? Are the foreign margins the same as in the United States?

Clark: The prospects for growth in all overseas markets are driven by GDP. Currently, we think Europe will continue to lag the United States in growth rates and in production facilities coming online. In the Far East, prospects for growth remain strong. In fact, as noted previously, growth has exceeded planned capacity additions for most of the products discussed. The one product looking relatively sensitive because of new facilities overseas is PVC. In general, growth rates in the

United States should fall between the growth rates of the Far East and those of Europe.

Even though Europe has primarily a naphtha-based system, European margins for most ethylene-derived products are weaker than those in the United States. This weakness stems from the surge in U.S. exports following the dramatic increase in U.S. capacity without a corresponding increase in consumption. Thus, margins in Europe have been depressed, and the results at many of the European chemical companies, including the integrated oil companies and their chemical segments, are worse than those in the United States.

The Oil Services Industry—Understanding Its Distinguishing Characteristics

James L. Carroll, CFA
Managing Director
PaineWebber, Inc.

Although oil and gas prices influence the stock performance of oil services industry firms from a near-term perspective, earnings and earnings momentum drive long-term stock movement.

Winston Churchill said war and politics are quite different. In war, you are killed only once; in politics, you can be killed many times. Oil services stocks are like politics. Portfolio managers generally view these stocks as deep cyclicals, so the timing of buying and selling is essential. This presentation focuses on four broad areas: what moves oil services stocks, what moves the fundamentals, forecasting earnings, and valuation and stock selection.

Oil Services Stock Movements

The performance of oil services stocks is closely related to changes in oil prices and oil price expectations. **Figure 1** illustrates how these stocks have moved slightly ahead of oil prices since 1979. The refiner's acquisition cost of crude is a good proxy for oil prices, because that index has the longest available data history.

The relatively close trading pattern with respect to natural gas is a fairly new development. **Figure 2** shows that when natural gas prices (quoted in dollars per million British thermal units [Btus]) had not yet been fully deregulated in the mid- to late 1980s, the group of oil services stocks led natural gas prices. Since full deregulation in 1989, the group has traded closely with natural gas spot prices. The steep increase in natural gas prices, which began in March 1992, pushed oil services stocks higher. In late March, these stocks were discounting a natural gas composite spot price close to $1 a thousand cubic feet (mcf). Although spot gas prices have improved recently, oil services stocks have lagged somewhat.

From a long-term perspective, earnings are important for this group. **Figure 3** illustrates Schlumberger's stock price plotted against earnings during a period when oil prices were not increasing, a period of $3-a-barrel oil and natural gas prices of 17 cents an mcf. The price of oil was easy to determine, because it never changed; it was set by the Texas Railroad Commission. During this period of oil and gas price stability, the stock closely followed earnings on a long-term basis. **Figure 4** shows that even after 1973, a period of great oil price volatility, the stock price continued to follow earnings closely. The exception was 1986 and 1987, when Schlumberger had a net loss because of the write-off and eventual divestiture of Fairchild Corporation's semiconductor operations.

During the past 20 years, oil services stocks have moved well in advance of the fundamentals. After the peak of the last price cycle in late 1980, for example, stock prices began to decline in sympathy with the oil price break in 1981. Earnings continued to rise until the first quarter of 1982, however. The key point, then, is that oil and gas prices move oil services stocks from a near-term perspective; earnings and earnings momentum move the stocks from a longer term perspective.

The connection between what moves the stocks and what moves the fundamentals is the worldwide rig count. As shown in **Figure 5**, the connection between the rig count and oil prices was tight until the past few years, when oil prices drifted higher and the rig count drifted lower in reflection of the steep decline in U.S. natural gas prices. The U.S. rig count has moved more with natural gas prices since full natural gas price deregulation in the late 1980s.

Figure 1. Refiner's Acquisition Cost and S&P Oil Well Equipment and Services Index, 1979–92

Sources: Energy Information Administration; Platt's *Oilgram News*; Standard & Poor's; PaineWebber estimates.

Figure 2. U.S. Natural Gas Spot Prices and S&P Oil Well Equipment and Services Index, 1985–92

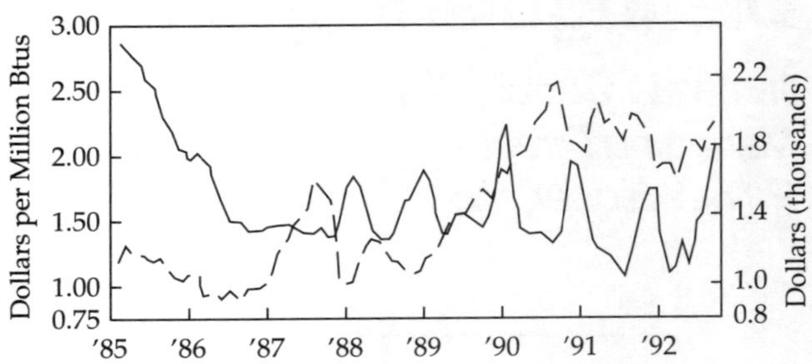

Sources: Natural Gas Week; Standard & Poor's.

What Moves the Fundamentals

In addition to oil and gas prices, two basic factors affect rig activity: the economics of exploration, and the availability of existing prospects. The major economic factors that affect the industry are new technology, international drilling activity, government policy, and the natural gas market.

New Technology

The costs of finding oil have been declining. **Figure 6** shows U.S. finding costs drifting from a high of almost $10 a barrel in 1986 to $6 a barrel in 1990 and then rising modestly in 1991 to about $7.50 a barrel. Because new oil field technology is reducing finding costs, this number should continue to move down. The rule of thumb is that a producer's return will typically exceed a threshold investment rate of 12 to 15 percent when the finding costs equal half the price of oil. For example, for a finding cost of $7.50 a barrel, a threshold oil price of about $15 a barrel would be needed for a marginal return. Non-U.S. finding costs are more volatile than U.S. costs, but they have also been falling since 1988. The volatility is mainly attributable to the large capital commitments required on frontier locations and the long lead times necessary to establish production.

Finding costs are falling because of new technology. Few analysts understand the role of technology. Many believe new technology means lower revenues and earnings to oil services companies because the companies do not need to drill as much as previously. In fact, improved technology reduces threshold oil prices or threshold structure sizes, so marginal fields become more profitable to develop. As a result, over time, more technology leads to greater revenues and earnings, because oil

Figure 3. Schlumberger Earnings and Stock Prices, 1962–74

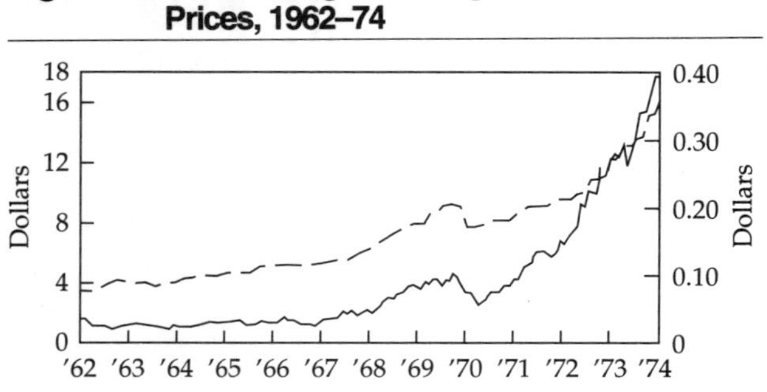

Source: PaineWebber.

Figure 4. Schlumberger Earnings and Stock Prices, 1973–92

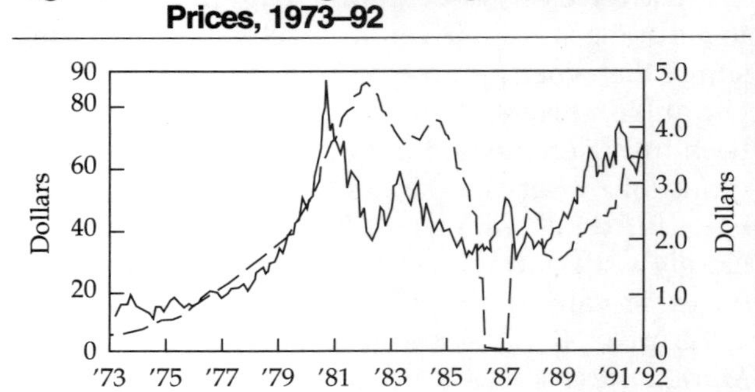

Source: PaineWebber.

Figure 5. Refiner's Acquisition Cost and Worldwide Rig Count, 1979–3Q'92

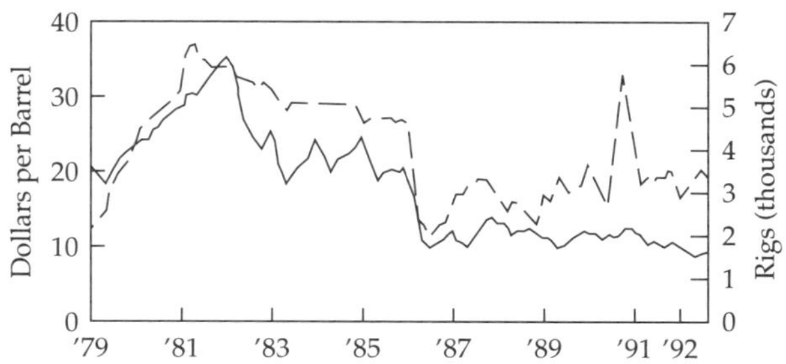

Sources: Energy Information Administration; Platt's *Oilgram News*; PaineWebber estimates; Baker Hughes.

Figure 6. U.S. Finding Costs, 1986–91

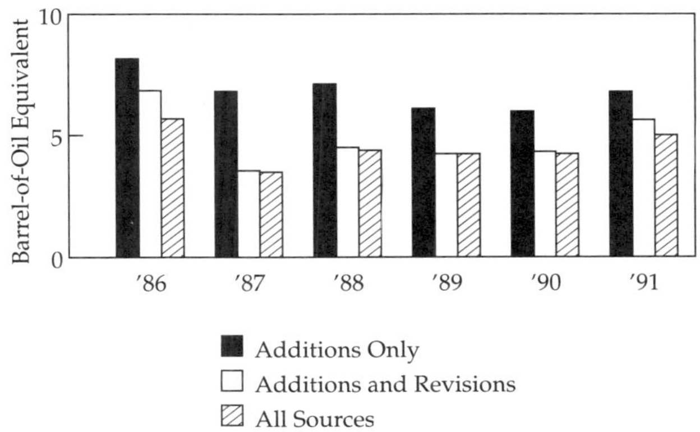

Source: Arthur Andersen & Co.

companies' investment returns rise as finding costs fall. More rigs are not necessarily running, but more exploratory and development services are used, and overall oil field revenues rise.

Table 1 provides information on new technologies used in the oil services industry and lists the companies likely to benefit from them. These technologies can be classified into three broad categories: exploration and drilling; production and field development; and workover and safety and environment. The major new product in exploration and drilling is basin modeling, which is the aggregation of core sampling data, seismic data, and wireline data from many different scientific sources to build a reservoir profile early in the drilling process. Better modeling methods improve success ratios. Currently, one out of six exploratory wildcat wells is successful, on average; where basin modeling has been used extensively, that ratio has dropped to one out of three.

International Drilling Activity

Three factors are important indicators for the foreign market: international seismic crew counts, acreage under license, and new concessions. The number of seismic crews is a leading indicator for drilling and also provides an idea of what is happening in the industry over time. As seen in **Figure 7**, the international seismic crew count rose sharply following the end of the Persian Gulf War (February 1991). Various countries granted the oil industry about 75 new concessions to develop properties.

Although the international rig count fell in 1992, acreage under license has been fairly stable in recent years, as illustrated in **Figure 8**. Acreage under license is a leading indicator of international drilling activity. The recent stability of acreage under license suggests that the downward decline in the international rig count is temporary.

Many new concessions have been granted or are under discussion in Albania, Argentina, Brazil, and Russia, primarily because these countries have no oil field technology, need hard currency, and are seeking access to Western markets. These concessions will have a positive long-term impact on international drilling, because entry into these areas now will allow drilling in places that have been ignored for years.

Government Policy

Tax incentives, tax preference items, and royalties and royalty moratoriums can also have a big impact on drilling. In the United States, a major incentive expired at the end of 1992—Section 29 tax credits. As for tax preference items, North Sea exploration has been heavily driven by the deductibility of most drilling expenses. A royalty moratorium was recently passed in Canada, which should result in more drilling there. Generally, government policy outside the United States is quite favorable toward drilling, probably more favorable than it has been for more than 20 years.

Figure 7. International Seismic Crew Count, 1988–3Q'92

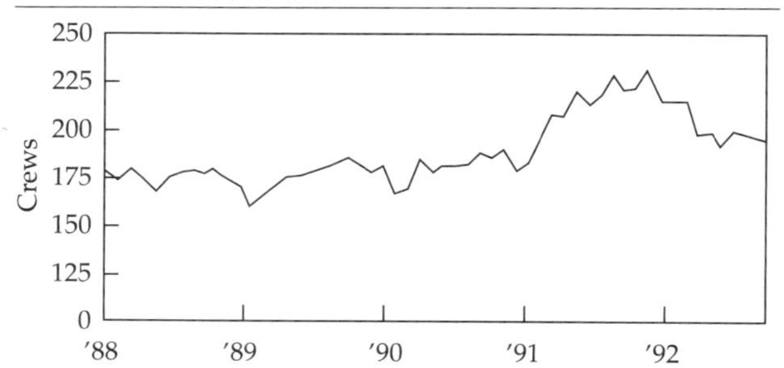

Source: Society of Exploration Geophysicists.

Table 1. New Technologies

Technology	Objective	Methods and Techniques	Companies Poised to Benefit
Exploration	Improve success ratio for exploratory or wildcat drilling	3-D seismics; integrated geological and geophysical approaches; basin modeling	SLB, Litton/DI, DGC, LMRK
Drilling	Reduce cost by 30 percent; improve safety performance	Smart drilling; PDC bits; oil-base mud; new line automation; MWD; instrumentation	SLB, VRC
Production	Improve oil recoverability from 30 percent average	Horizontal wells; geological reservoir description and simulation; enhanced oil recovery; completion technology; fluid monitoring	BHI, BRC, SLB
Offshore fields	Reduce investment costs	Platform optimization (location, weight, and number); horizontal/extended reach wells; subsea completions; multiphase production logging	CBE
Workover	Reduce cost by 50 percent	Intelligent coiled tubing; multiphase production logging	BHI, HAL, SLB, NWELF, BJS
Safety and environment	Minimize pollution/accident risks and exposure	Drilling fluids; continuous mix processes; environmentally compatible chemicals; automation of rigs and platforms; elimination of radioactive sources	NPRS

Sources: Schlumberger; PaineWebber.

Key:

BHI	= Baker Hughes	Litton/DI	= Litton/Dresser Industries
BJS	= BJ Services	LMRK	= Landmark Graphics
BRC	= Baroid Corp.	NPRS	= New Park Resources
CBE	= Cooper Industries	NWELF	= Nowsco Well Service
DGC	= Digicon Inc.	SLB	= Schlumberger
HAL	= Halliburton Co.	VRC	= Varco International

Natural Gas Market

Fundamentally, the natural gas market is critical to the oil services industry for three reasons: U.S. gas-related drilling accounts for 70 percent of the domestic industry's profits; domestic capacity utilization typically depends on U.S. gas drilling; and the domestic market sets worldwide product and service pricing.

The natural gas market has had many problems, including very weak demand, although as **Table 2** indicates, demand has been rising in the past few years. Demand peaked at 22 trillion cubic feet (tcf) in 1973 and dropped by 6 tcf by the mid-1980s, mainly because of the Fuel Use Act of 1978 passed during the Carter administration. Demand has been improving since 1990 but is still less than it was 12 years ago.

Supply additions in the gas market have been a matter of some confusion. **Table 3** shows the percentage change in various measures of the size of the gas market from 1980 to 1992. Reserve additions per gas well completion increased dramatically from a factor of 0.80 in 1985 to an estimated 1.33 in 1992. Many observers have said this increase is a reflection of oil field technology. In my opinion, however, it reflects a major change in the approach to drilling for gas; producers are now looking for large gas structures in the U.S. Gulf of Mexico rather than pursuing smaller, land-based gas drilling. As structure size has risen, reserve additions per gas well completion have also risen.

In 1985, the Reagan administration introduced areawide leasing in the Gulf of Mexico, which opened huge structures not previously available for exploration. After mid-1985, an average of about 120 rigs a year were running in the Gulf of Mexico, and the area was providing a large percentage of our total reserve additions. Nevertheless, drilling in the Gulf of Mexico declined through most of 1991 and collapsed in early 1992. Only 50 rigs were running in the Gulf of Mexico in June 1992. This level of drilling will not come close to replacing the production lost in 1991 and 1992. For example, in 1992, we replaced about 7.5 tcf of gas production before revisions, and that number will likely fall in 1993.

The natural gas surplus in 1991 was estimated at 5 tcf. As **Table 4** shows, the surplus fell by half in 1992 alone because of lack of drilling in the Gulf of Mexico. The end of the Section 29 tax credit in 1992 will result in reduced gas drilling, which will put pressure on new reserve additions. Thus, the gas market is balancing somewhat. A surplus remains, but it is not nearly as large as it once was (it is now running about 1 tcf), and supply is much closer to

Figure 8. Acreage under License by Region, 1987–91

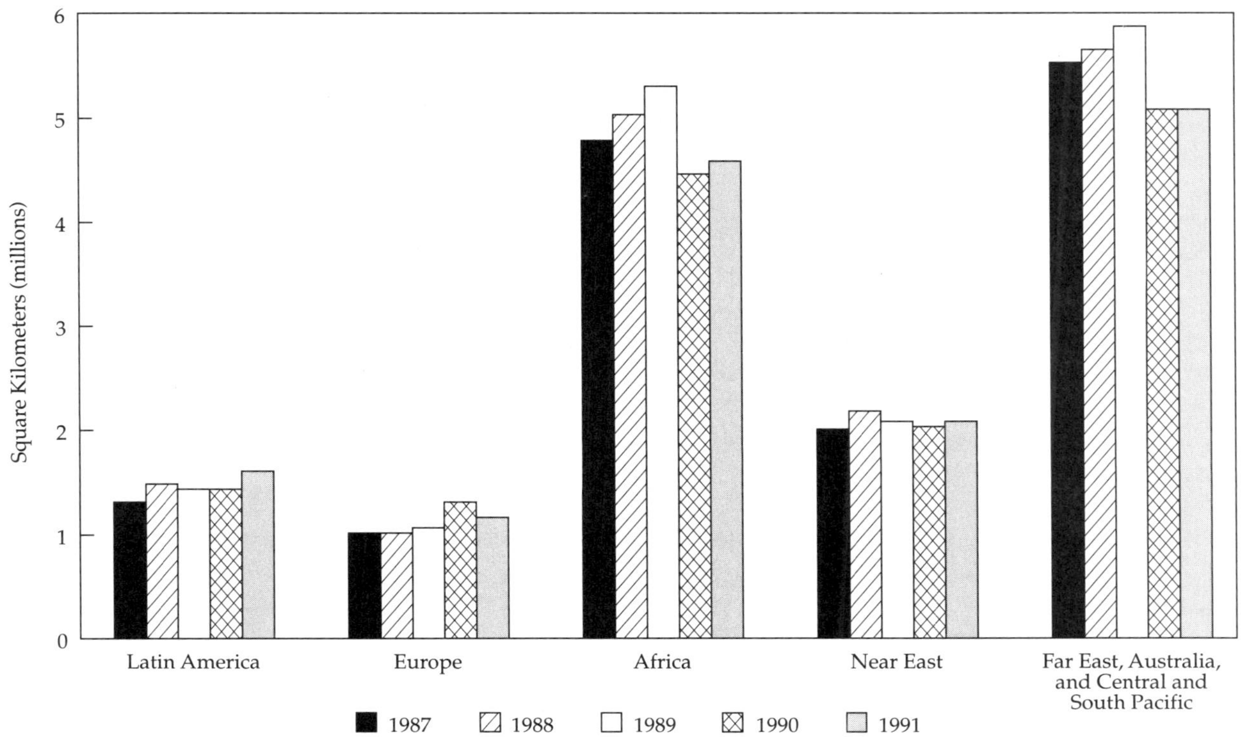

Source: Petroconsultants.

demand than it has been in a long time. With gas drilling still weak and sparse reserve additions, this trend should continue.

Forecasting Earnings

The first step in forecasting earnings is forecasting industry revenue. Wellhead revenue is the ultimate source of domestic revenue for oil services companies. A wellhead revenue model is shown in **Table 5**. Plowback is probably the most critical variable in this model. It reflects the percentage of wellhead revenue dedicated to reserve replacement. Over a 40-year period, plowback has averaged about 20 percent; it has been as high as 33 percent (in 1981) and is a low 9.5 percent today. Plowback should increase to the midteens as more remedial work occurs on U.S. fields.

Plowback drives the drilling expenditure forecast, which is essentially nominal revenue for oil services companies. From drilling expenditures, we

Table 2. Natural Gas Demand, 1973–92[a]

	Natural Gas Demand by Segment (million cubic feet)						
Year	Residential	Commercial	Industrial	Electric Utilities	Other	Total	Percent Change
1973	4,879	2,597	8,689	3,660	2,224	22,049	
1980	4,752	2,611	7,172	3,682	1,660	19,877	–9.9%
1985	4,433	2,318	5,901	3,044	1,470	17,166	–13.6
1986	4,314	2,430	5,579	2,602	1,408	16,333	–4.9
1987	4,315	2,670	5,953	2,844	1,668	17,450	6.8
1988	4,630	2,719	6,383	2,636	1,710	18,078	3.6
1989	4,781	2,718	6,816	2,787	1,700	18,802	4.0
1990	4,391	2,680	6,970	2,787	1,896	18,724	-0.4
1991	4,565	2,781	7,321	2,788	1,929	19,384	3.5
1992[a]	4,600	2,780	7,600	2,863	1,929	19,772	2.0

Source: U.S. Department of Energy.

[a]Estimate.

Table 3. Gas Drilling and Reserves, 1980–92[a]
(million cubic feet, except percents)

Year	Total Gas Well Com-pletions	Percent Change	Gas Reserve Additions	Percent Change	Reserve Addition per Gas Well Com-pletion[b]	Percent Change	Average Number of Rigs Drilling for Gas	Percent Change	Reserve Addition per Gas Rig[c]	Percent Change
1980	15,225		14,473		0.95		NAv	NAv	NAv	NAv
1981	17,533	15.2%	17,220	19.0%	0.98	3.3%	NAv	NAv	NAv	NAv
1982	17,052	–2.7	14,455	–16.1	0.85	–13.7	NAv	NAv	NAv	NAv
1983	12,919	–24.2	11,448	–20.8	0.89	4.5	NAv	NAv	NAv	NAv
1984	14,983	16.0	13,521	18.1	0.90	1.8	NAv	NAv	NAv	NAv
1985	13,854	–7.5	11,128	–17.7	0.80	–11.0	NAv	NAv	NAv	NAv
1986	7,351	–46.9	8,935	–19.7	1.22	51.3	NAv	NAv	NAv	NAv
1987	7,255	–1.3	7,175	–19.7	0.99	–18.6	388	—	18.49	—
1988	7,820	7.8	10,350	44.3	1.32	33.8	351	–9.5%	29.49	59.5%
1989	8,830	12.9	10,032	–3.1	1.14	–14.2	401	14.2	25.02	–15.2
1990	9,474	7.3	12,368	23.3	1.31	14.9	465	16.0	26.60	6.3
1991	6,158	–35.0	7,564	–38.8	1.22	12.0	345	–25.8	26.09	–1.9
1992[a]	6,000	–2.6	8,000	5.8	1.33	9.0	300	–13.0	26.60	2.0

Sources: American Petroleum Institute; U.S. Department of Energy; PaineWebber; Baker Hughes.

Note: NAv = data not available.

[a]Estimate.
[b]Gas reserve additions divided by total gas well completions.
[c]Gas reserve additions divided by average number of rigs drilling for gas.

forecast some of the macro factors that are influential for oil services revenues: rig count, footage, and total wells drilled, which are listed in **Table 6.**

Forecasting revenues for the oil services industry involves certain problems. The most significant is budget surveys (the total capital budgets of the oil and gas industry). Oil services companies live and die by budget surveys. One reason oil services stocks are down this week is because the American Petroleum Institute just announced that its members will not spend more money next year than this year. Investors decided that announcement meant these stocks are in trouble. Budget surveys are never accurate, however. Companies either spend more or less, never exactly what they tell you they will spend. Quite simply, they spend more if business is better and less if it is worse.

Another problem is the survey itself. Every survey looks at the major U.S. companies, the independents, and some smaller companies. Of the top 25 oil companies in the world, however, 14 are national and only 11 are major oil companies. The national

Table 4. Natural Gas Reserves and Surplus, 1981–93[a]
(trillion cubic feet, except as noted)

Year	Total U.S. Reserves January 1[b]	Total U.S. Reserves Replaced	Dry Gas Production	Net Revision and Adjust-ment	Dry Gas Delivered[c]	Delivery Rate	Estimated Surplus	Surplus as Percent of Delivery	U.S. Lower 48 Reserve Life (years)	Average Wellhead Price ($/tcf)
1981	199.0	17.2	18.7	4.2	21.3	10.7%	2.6	12.0%	10.6	$1.98
1983	201.5	11.5	15.8	3.1	21.0	10.4	5.2	24.8	12.8	2.59
1985	197.5	11.1	16.0	0.8	21.0	10.6	5.0	23.9	12.4	2.51
1987	191.6	7.2	16.1	4.6	20.0	10.4	39.0	19.5	11.9	1.67
1988	187.2	10.4	16.7	–12.9	20.0	10.7	3.3	16.6	11.2	1.69
1989	168.0	10.0	17.0	6.0	21.0	12.5	4.0	19.1	9.9	1.69
1990	167.1	12.4	17.2	7.1	22.0	13.2	5.0	22.7	10.1	1.71
1991	169.3	9.0	17.2	7.4	22.0	13.0	5.0	22.7	9.8	1.59
1992[a]	167.0	9.0	17.3	6.0	20.0	12.1	2.7	13.5	9.7	1.50
1993[a]	162.7	10.0	17.4	4.0	19.0	11.8	1.6	8.4	9.4	1.60

Sources: U.S. Department of Energy; PaineWebber.

[a]Estimate.
[b]Excludes Alaska; includes nonassociated gas.
[c]Maximum daily production capability.

Table 5. PaineWebber Wellhead Revenue Model, 1981–97[a]

	Crude Oil Marketed Production				Natural Gas Marketed Production										
Year	Total Crude Production (thousands of barrels)	Percent Change	Net Realization ($/barrel)	Crude Wellhead Revenue ($billions)	Total Gas Production (tcf)	Percent Change	Average Gas Realization ($/mcf)	Gas Wellhead Revenue ($billions)	Total Crude/Gas Wellhead Revenue ($billions)	Plowback Assumption	U.S. Drilling Expenditure ($billions)	Percent Change	U.S. Rig Count	Percent Change	Drilling-Expenditure per Rig ($millions)
1981	8,572	—	$23.36	$73.09	19.18	NAv	$1.98	$37.98	$111.07	33.0%	$36.6	NAv	3,970	NAv	$ 9.09
1982	8,649	0.9%	22.58	71.28	17.82	–7.1%	2.46	43.84	115.12	34.2	39.4	7.6%	3,105	–21.8%	12.53
1983	8,688	0.5	22.21	70.43	16.09	–9.7	2.59	41.68	112.11	22.4	25.1	–36.2	2,232	–28.1	11.07
1984	8,879	2.2	22.52	72.98	17.47	8.5	2.66	46.46	119.44	21.1	25.2	0.4	2,428	8.8	10.82
1985	8,971	1.0	21.95	71.87	16.45	–5.8	2.51	41.30	113.17	20.9	23.7	–6.1	1,980	–18.5	11.97
1986	8,680	–3.2	12.51	39.63	16.06	–2.4	1.94	31.15	70.79	19.2	13.6	–42.5	964	–51.3	14.11
1987	8,349	–3.8	15.40	46.93	16.62	3.5	1.67	27.76	74.69	12.3	9.2	–32.4	937	–2.8	9.82
1988	8,140	–2.5	12.58	37.38	17.10	2.9	1.69	28.90	66.28	15.8	10.5	14.0	937	0.0	11.21
1989	7,613	–6.5	15.86	44.07	17.31	1.2	1.69	29.26	73.33	13.2	9.7	–7.6	1,008	7.6	9.62
1990	7,355	–3.4	20.03	53.77	17.81	2.9	1.71	30.46	84.23	13.0	10.9	13.1	870	–13.7	12.53
1991	7,373	0.2	16.50	44.40	17.87	0.3	1.59	28.41	72.81	12.8	9.3	–14.9	862	–0.9	10.79
1992[a]	7,300	–1.0	18.00	47.96	18.00	0.7	1.50	27.00	74.96	9.5	7.1	–24.0	675	–21.7	10.50
1993[a]	7,100	–2.7	18.00	46.65	18.00	0.0	1.65	29.70	76.35	10.5	8.0	13.2	750	11.2	10.69
1994[a]	6,900	–2.8	19.00	47.85	17.50	–2.8	1.75	30.63	78.48	12.4	9.7	21.5	910	21.4	10.70
1995[a]	6,700	–2.9	20.00	48.91	17.00	–2.9	1.90	32.30	81.21	13.8	11.2	14.7	1,015	11.5	11.00
1996[a]	6,500	–3.0	21.00	49.82	17.00	0.0	2.05	34.85	84.67	14.2	12.0	7.7	1,075	5.9	11.18
1997[a]	6,300	–3.1	22.00	50.59	16.50	–2.9	2.15	35.48	86.06	15.0	12.9	7.4	1,125	4.6	11.48

	Plowback Assumption	Drilling Expenditure per Rig ($millions)
1970–92[a] average	20.3%	$7.87
1981–92[a] average	18.9	11.17

Sources: Energy Information Administration; PaineWebber estimates.

Note: NAv = data not available.

[a]Estimate.

Table 6. U.S. Drilling Summary and Forecast, 1985–97[a]

Year	Average Rig Count	Percent Change	Footage (millions)	Percent Change	Total Wells Drilled	Percent Change	Drilling Expenditures	Percent Change
1985	1,980	–18.5%	314.5	–10.3%	67.8	–12.2%	$23.7	–10.3%
1986	964	–51.3	176.3	–43.9	37.2	–45.2	13.6	–42.8
1987	937	–2.8	157.4	–10.7	33.0	–11.1	9.2	–31.8
1988	937	0.0	150.2	–4.5	29.7	–10.0	10.5	14.2
1989	870	–7.2	131.5	–12.5	26.7	–10.2	9.7	–8.3
1990	1,008	15.9	143.8	9.4	28.5	6.8	10.9	13.1
1991	862	–14.5	133.2	–7.4	26.6	–6.8	9.3	–15.0
1992[a]	675	–21.7	105.0	–21.1	20.7	–22.0	7.1	–26.6
1993[a]	750	11.2	115.0	9.5	22.8	10.0	8.0	13.2
1994[a]	910	21.4	135.0	17.4	26.8	17.5	9.7	21.5
1995[a]	1,015	11.5	150.0	11.1	29.6	10.5	11.2	14.7
1996[a]	1,075	5.9	162.0	8.0	32.0	8.0	12.0	7.7
1997[a]	1,125	4.6	175.0	8.0	34.4	7.5	12.9	7.4

Sources: World Oil; American Petroleum Institute; Baker Hughes; PaineWebber estimates.

[a]Estimate.

oil companies have far surpassed the majors in spending money. We estimate that the national oil companies spend about 35 percent of total worldwide exploratory dollars. Saudi Aramco's budget is twice that of Exxon Corporation. The National Iranian Oil Company is quickly moving up toward the Saudi level of spending. Surveying these companies is a waste of time, however, because they exaggerate their numbers. As a result, accurate information is extremely difficult to obtain. Based on the service companies we talk to, the national oil companies will spend more money in 1993 than in 1992, but the surveys do not reflect these increases.

Oil prices are the major factor that determines international activity from a long-term perspective. The foreign market is tricky to forecast because wellhead revenue analysis cannot be done. Many countries do not have wellhead revenue. Because they are investing for future wellhead revenues, exploration and drilling tend to be prospect driven. As an indicator, we look at how many dollars the major companies are allocating to international markets. We also try to talk to the national oil companies.

The operating earnings model for a specific company is keyed off the industry revenue model. We forecast revenues for each business line separately: wireline (separated into North American and foreign operations), seismic, contract drilling, well stimulation, and so forth. After forecasting revenues, we forecast operating earnings. Major costs are labor, depreciation, and research and development. When operations are profitable, incremental margins become critical to forecasting earnings. Early in a cycle, when companies are running well below capacity, incremental margins average 30–40 percent. Near the peak of the cycle, however, incremental margins will decline to the midteens.

Trough, peak, and normalized operating profit margins for three companies are shown in **Table 7.** We have attempted to determine normalized earn-

Table 7. Review of Company Operating Profit Margins and Return on Equity
(percent, except rig count)

	1973 Trough	1981 Peak	1986 Trough	Normal	1992[a]	1993[a]
Operating profit margins						
Baker Hughes	9.1%	21.9%	–51.7%	16.0%	9.6%	12.8%
Halliburton Co.	15.2	24.5	–19.7	14.6	4.1	7.7
Schlumberger	29.8	44.9	0.3	21.3	15.1	16.9
Return on equity						
Baker Hughes	12.6	26.8	NM	15.4	6.3	10.3
Halliburton Co.	15.2	3.9	NM	15.9	3.5	7.5
Schlumberger	13.2	29.9	NM	23.5	17.8	18.5
Worldwide rig count	2,156	5,625	2,215	—	1,669	1,840

Source: PaineWebber.

Note: NM = not a meaningful figure.

[a]Estimate.

ings estimates for these companies; this midcycle level of earnings is critical for the valuation of the stocks.

The oil services companies vary in growth of revenues per rig. This measure is important when forecasting revenue, because it measures how well a company's products are accepted by the marketplace. Schlumberger's growth rate of revenue per rig has been almost 20 percent a year for the past several years. At the opposite end, Halliburton Company's growth rate has been only 5.4 percent.

As with any industry, the oil services group has several accounting practices of which to be aware. The biggest differences in accounting are depreciation methods. Some oil services companies do not depreciate idle equipment and, as a result, overstate earnings. Liquidity requirements in loan covenants and tax-rate manipulations are always warning signals. Asset write-downs are another method companies use to increase earnings through accounting. Halliburton announced a considerable asset write-down, for example, that will add about 15 cents to its earnings in 1993. Revenue recognition can also be an accounting pitfall. Does a company use completed contracts or the percentage-of-completion method in reporting revenues?

Valuation

Oil services stocks tend to trade together. When the tide goes out, all the ships go with it. It is a deep cyclical group, so timing is critical. Stock prices closely follow discounted cash flows, as seen in **Figure 9.** At the bottom of the cycle, the stocks trade at about 55 percent of their discounted cash flow values; at the top of the cycle, they trade at more than 100 percent. The most recent peak in value was when Saddam Hussein invaded Kuwait. The low points are typically oil price panics. Currently, the group is trading at about 70 percent of discounted cash flow value, which reflects a major correction in the past few weeks.

Figure 10 illustrates how these same stocks have traded on market multiples. Generally, they have sold at a premium to the market over long periods because of the nature of this industry: It can reduce costs, it can downsize, and it can be profitable at almost any level of activity or any level of oil or gas price.

Picking Winners

Certain criteria are very helpful in picking winners in the oil services industry:

- *Companies with debt of less than 30 percent of total capital.* A strong balance sheet is a priority because it yields a much higher valuation premium than other companies will achieve.
- *Companies with a strong franchise in either a critical product line or a niche service.* In this industry, the cost of the service is not high relative to what is at stake in the ground. For that reason, reputation and product quality are important, and this market does not have a lot of foreign competition: Too much is at risk to take a chance on an unknown and/or less expensive foreign drill bit.
- *Companies that are low-cost producers.* This criterion has become important because the business has recently become much more competitive.
- *Companies with products that reduce finding*

Figure 9. Stock Price as a Percentage of Discounted Cash Flow— Four Oil Services Companies, 1985–3Q'92

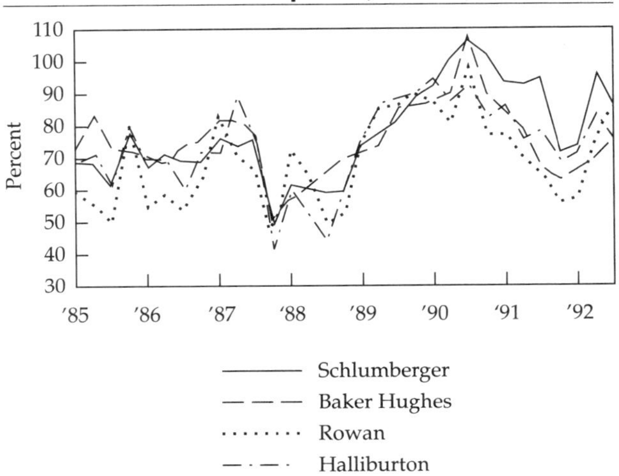

Source: PaineWebber.

Figure 10. Price-to-Earnings Ratios, Four Oil Services Companies, 1970–92

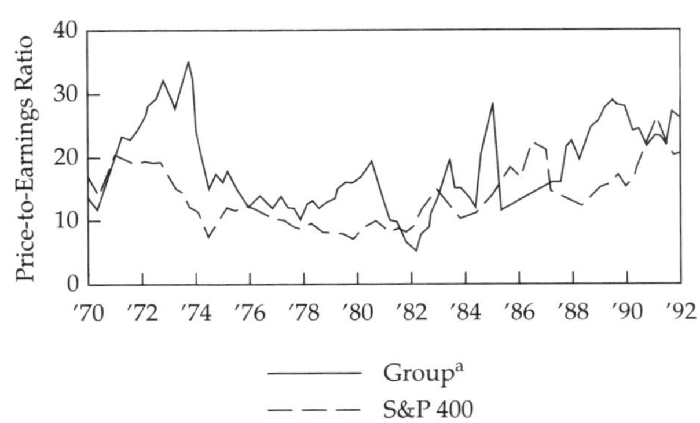

[a]Includes Schlumberger, Baker Hughes, Rowan Companies, and Halliburton.

Source: PaineWebber.

Note: 1986 is not a meaningful figure for group.

costs or the risk in exploration.

- *Companies in which management has a meaningful stake in the company.* Many people are still riding on the past successes of these companies. They are not managing the companies the way they should be for the 1990s.
- *Companies in which outside directors have a substantial role.*
- *Companies with rising research and development commitments and relationships with the research-intensive majors.*

Conclusion

Oil services stocks are driven by oil price changes and gas price changes early and late in cycles. The move in these stocks in 1992 has been related to higher oil and gas prices, particularly gas prices. The pattern is typical of what happens to these stocks early in a cycle: They move up when earnings are falling. The fundamentals still depend on industry economics, which are getting better because of greater international access and because new technology is lowering finding costs.

Picking winners is a process of trying to focus on companies that are well positioned for those markets, have strong balance sheets, and offer value-added products. But remember, oil services is a cyclical industry; do not forget to sell when valuation targets have been achieved.

Question and Answer Session

James L. Carroll, CFA

Question: How reliable a measure of potential future revenues is the seismic crew count?

Carroll: The seismic crew count is an indicator of exploration activity, particularly outside the U.S. market. In fact, the international seismic crew count in 1992 was the only indicator that the international rig count was headed lower. One problem with that indicator is the shelf life of seismic data. Companies usually hire a seismic crew before they drill because they need to know where to drill. If they have a lot of data on the shelf, then they may not have to use a seismic crew. Three-dimensional (3-D) seismic technology has changed that, however. Companies drilling internationally use 3-D seismic surveys, and those are not sitting on a shelf. In the domestic market, seismic crew count is not as good an indicator of exploration activity, because most operators are still using older two-dimensional data. But this is beginning to change, particularly in the U.S. Gulf of Mexico.

Question: Who are the technological leaders in the 3-D seismic market?

Carroll: The 3-D market has three segments: companies that acquire the data, companies that process the data, and companies that make systems for the acquisition and processing of the data. The largest players in the 3-D market are Western Atlas, a division of Litton and Dresser, and Geco-Prakla, a division of Schlumberger. They are the largest acquirers and processors of data. Many small companies manufacture systems. Input/Output makes a 24-bit land data acquisition system that is highly useful in the 3-D market. Landmark Graphics makes work stations that allow a geophysicist or geologist to take the raw seismic data from a 3-D survey and turn it into something useful. Digicon is the only pure play in the marine 3-D acquisition market. With only nine vessels, it is a smaller company than Geco-Prakla, which has 28.

Question: What has been the effect of the Section 29 tax credit?

Carroll: Section 29 credits are responsible for about 230 out of 853 rigs in the Baker Hughes rig count in the United States. In fact, probably about 400 rigs are working on that credit. The balance of them are not in the count because they are not drilling rigs; they are cable rigs or water well rigs. An operator can drill two or three 1,300- or 1,400-foot Section 29 wells a week, so they do not provide much revenue for the mainline service companies. They stimulate demand for Halliburton, BJ Services, or Western, but they do not use many drill bits, and they do not use wireline services.

Now that the Section 29 credit has expired, there will be a fairly steep drop in the U.S. rig count. Everybody spent budget money in anticipation of this credit expiration at year-end 1992. But many of these wells can produce economically with gas prices below $2 an mcf. We believe that U.S. gas drilling will recover in the second half of 1992, despite a very steep decline immediately after the credit expired. This is directly related to higher U.S. natural gas prices.

Question: Please comment on the effect of the changes in the alternative minimum tax (AMT).

Carroll: The AMT was repealed in the energy bill, which provided producers with about $1.1 billion in incremental cash flow. Senator Don Nickles (R, Okla.) conducted a study indicating that 1992 AMT alone cost the United States more than 200 drilling rigs, because producers were forced to pay a minimum tax regardless of profitability. My own work shows that between 50 and 100 rigs could go back to work as a result of that credit. But the U.S. rig count was greatly depressed in 1992 for many reasons. Gas prices at $1.50 or $2 an mcf will stimulate higher demand next year for a more normal level of gas-related drilling.

Question: Which oil services companies meet the criterion of less than 30 percent debt to capital?

Carroll: Most of them do. I cover about 20 companies, and about 16 have debt to total capital less than 30 percent. Many companies have no debt. Schlumberger, for example, carries a $1 billion to $2 billion cash balance. It has about $800 million in debt, but on a net basis, it is debt free. Dresser is debt free, Halliburton's ratio of debt to total capital is well below 30 percent, and Baker is close to that number. The bulk of the companies I follow have very little financial leverage. They cannot afford to have much, given the volatility of the markets.

The Art of the Oil and Gas Interview

Arthur L. Smith, CFA
Chairman and CEO
John S. Herold, Inc.

Edd Grigsby
Manager of Investor Relations
Phillips Petroleum Company

This presentation consists of comments from the perspective of the analyst interviewer (Smith), comments from the perspective of the investor relations interviewee of a major oil company (Grigsby), and a mock interview.

The Interviewer (Smith)

The analyst's purpose in an interview is to obtain as much information and perspective on a company as possible in a short period of time. Anyone can retrieve data on a screen or read *Value Line* or *Standard & Poor's* tear sheets. Anyone can read an annual report, although few actually do. Great analysts get behind the numbers and understand the corporate soul.

I feel strongly that interviews are an important part of solid and necessary due diligence. I am amazed by the extent to which investing in our industry is black box investing—that is, based solely on charts, statistics, or new numbers. Such an approach is equivalent to making an investment decision to buy a $2 million house or $20 million factory without inspecting the property, checking out the architect, or seeing how the current owner maintains and operates the infrastructure.

Analysts can learn a great deal from observing the offices of an oil company. In years past, I could tell oil companies were in trouble when I saw more investment in artwork and thick pile carpets than in the field. Today, everyone prefers lean-and-mean companies. When visiting executive offices, I am reassured to find money invested in technology, not furniture.

At my firm, interviews are conducted on three general levels. An interview with the oil company's investor relations person provides a general overview of the company and the data an analyst needs. An interview with the CEO gives the analyst an opportunity to ask about strategic issues and determine top management's vision for the company. Finally, the analyst may have an opportunity to interview divisional heads of operations. In the divisions, the analyst may find particular individuals with intensive knowledge about specific segments of the business.

When interviewing managers, analysts can benefit by heeding a simple list of do's and don'ts.

The Do's

Analysts should follow certain guidelines in the interview process:

- *Do your homework before the meeting*. Asking questions that are answered in the annual report is a waste of valuable time. Each direct visit an analyst makes to an oil company's headquarters costs at least $1,000, which includes the analyst's time away from the office, the time taken by the company's investor relations and corporate staff, and travel and entertainment expenses. Therefore, study the company's annual report and industry research reports for background information before the meeting. The company's history is also important. How have operating results fared over the long term and through industry crises?
- *Be punctual*. Analysts waste everyone's time when they arrive late. Do not try to schedule too many interviews in a short period. For example, in a one-day trip to the Dallas/Fort Worth area, given the geography of those cities, four interviews are

probably all an analyst can handle.

■ *Stay on track.* Time is precious, so an analyst must have an agenda and stick with it. Do not talk about the weather! Many times, when a company has only an hour to give, analysts wander around and never really get to the key issues. Interview questions should be well thought out, and the analyst should work from an outline. Asking about the interview objectives is helpful; they vary among companies.

■ *Be prepared to ask tough, penetrating questions, because the company often has tough, penetrating questions for the analyst.* Many analysts also find it valuable to ask the same question of different people in the company to see how well the message of management is being carried through the ranks.

■ *Ask for copies of Wall Street research on the company.* Many companies freely circulate research reports on their company to the buy- and sell-side analysts in the industry. Although my firm discourages widespread copying of research, analysts should use the tricks of the trade.

■ *Schedule a field trip.* When possible, see the company's operations firsthand and get to know the assets in which you are investing.

■ *Send management a copy of the completed research report.* As a courtesy, analysts should send a copy of the research report to management. This practice will help foster understanding and communication.

Beware if you feel comfortable with everything, because a herd mentality plagues the oil industry. Generally, everyone invests wildly in an exploration play, which leads to poor returns. Remember Fischer's Law: A conclusion is a place where you got tired of thinking.

The Don'ts

Analysts should avoid certain behavior in the interviewing process:

■ *Do not interrogate management in an adversarial manner.* Companies do not have to be helpful, open up, or even talk to analysts. Antagonizing management will only hinder the research. Some companies will not send a 10-K or annual report, so analysts have to buy the information from commercial vendors such as Disclosure.

■ *Do not get hung up on the numbers.* Overemphasis on number crunching is a common problem in the investment industry. One investor relations person said her biggest complaint about a meeting was that the analyst spent two hours asking for numbers she could not immediately provide. If the numbers are needed, the analyst should send a request in advance so the staff has time to dig them up. Then, analyst and managers will have time to talk about substantive issues during the visit.

■ *Do not ask for insider information.* This point is covered under the AIMR Code of Ethics and Standards of Professional Conduct, and AIMR members know what areas are appropriate to investigate. Do not ask about merger transactions, acquisitions, or any item about which the company clearly cannot comment.

■ *Do not talk too much.* Keep in mind the old saying: Good listeners are not only popular everywhere, but after awhile, they know something. Analysts should meet with top personnel, listen, ask questions, and find out as much as possible. They should not consider the interview an opportunity to give unsolicited advice. Some analysts simply forget the reason they are visiting management!

■ *Stick to important issues.* Analysts need to know how management conducts its business and makes its decisions, but they do not need to know all the technical details. Do not let the interview turn into a lesson on subsurface geography or geology.

■ *Do not chase other people's money in exploration plays.* Make sure management has its own money at stake. As a screening mechanism, I often use the question: Can the prospect be farmed out? If the answer is no, count me out.

■ *Do not fall in love with management.* Analysts must remain impartial and understanding. Determine where the company is going. Quarterly earnings are not the point; what is important is how the company is adapting to significant changes in the industry.

Good analysts always ask about the competitive advantages of the company, who the company considers its peers, and what the company believes it does better than the competition—even if the analysts believe they know the answers. As Bertrand Russell said, "In all affairs, it's a healthy thing now and then to hang a question mark on the things you have long taken for granted."

Final Reminders

Analysts should pay attention to what makes an oil company successful. In my view, the keys to success are solid capital structure and willingness to raise equity. No oil company managers think their stocks are fairly valued. Nonetheless, they must be willing to go to the market occasionally to keep their balance sheets in line. The company should be focused on areas of competitive advantage; geography, technology, and cost-efficiency are important. The company should have a good compensation structure that is clear to management and the outside world.

Successful companies also set high hurdles for performance. Canadian exploration and production companies, for example, disclose their targets in advance, and they hit them. They know their competitive environment. Although relatively few companies in this business today are adopting their competitors' good ideas, intense competitor intelligence gathering will increase in the future.

Analysts should also verify that management's numbers make sense. The oil industry tends to be driven by billion-barrel pipe dreams. Geologists can typically find 10 times as much oil on paper as the engineers can extract from the ground. Analysts should thus question any claims that are in the billions and trillions. An average gas field is 4–6 billion cubic feet. A trillion cubic feet of gas cannot exist under one section.

Managers and companies that have their act together have a good time. When analysts walk into the offices of a well-run oil company, they will notice that people are motivated and excited about their work.

The Interviewee (Grigsby)

Interviews vary, but the basic intent of the oil company is to communicate certain information clearly. In the first 10 minutes of a meeting with an analyst or portfolio manager, I do three things. First, I position Phillips Petroleum within the industry—who we are, what size we are, and how we stack up. Second, I explain our strategy for each significant business line. Our company is labeled a domestic integrated oil company and is involved in three or four basic businesses. We find and produce oil, refine it, and upgrade it into value-added products. Third, I highlight the new events in the industry that are having a significant impact on the company or, the significant events that deal specifically with Phillips. Then, I am ready for questions. This approach works well, because in a short time, the analyst gains our sense of the company, which may be completely different from the way the analyst sees us.

We have tried diligently to overcome one typical investor relations shortcoming; that is, an analyst should not have to ask us about any single historical number. The numbers should be on the analyst's desk. We try to be proactive in providing information and organizing it in such a way that analysts understand it. They should not have to go to too many different places in the annual report to tie capacity together with production, or revenue together with prices, and so forth. I rarely receive such questions from analysts, but when I do, it is an indication that we have fallen short of our goal.

Although the financial information about our company is easily communicated, the strategic plan takes longer to explain. I make sure the analyst understands that we have a very specific strategic plan for each major piece of the business. For example, the financial management plan is intended to increase our financial flexibility. Part of the plan is reducing debt. I show our debt position—where we have come from and where we are today. The analyst can then put that in perspective with regard to where we think we are going in the future.

The plan also calls for maintaining a competitive dividend. When talking with a portfolio manager, I explain how we develop our dividend policy within the company—what we mean by competitive yield or dividend, what our shareholder return objectives are, earnings and income payout, and so forth. The analyst should have no question about what we are trying to do.

After discussing our objectives, I explain what we have done for shareholders. Shareholder returns for the five years ending December 31, 1991, were 19.3 percent, compared with 14.2 percent for the S&P 500, and were higher than those of other major oil companies. We measure ourselves for incentive compensation based on shareholder returns.

Debt is a big issue for Phillips. Even portfolio managers who do not know our company well know that we are a small oil company and are highly leveraged. Our company announced an asset sale program in the fall of 1991, and we gave ourselves until the end of 1993 to sell $500 million worth of assets. When meeting with analysts, I update them on that program and on others, such as our program to improve cash flow.

Based on feedback from colleagues in other industries, I believe oil analysts are far better informed than analysts following other industries. Oil analysts, particularly on the sell side, can compete with anybody in their understanding of the industry and the depth of the research they perform.

Analysts ask about our competitive advantages and our position within the industry. Phillips differentiates itself within the industry on the basis of how successfully it replaces reserves, a key factor for an oil company. During the past five years, we have had the lowest finding and development costs of any oil company. In short, we have been successful in replacing reserves, and we are doing it cheaper than anyone else in the industry. That is important for analysts trying to figure out what will happen to the company in the future. In time, depreciation, depletion, and amortization charges will reflect these low finding and development costs.

Analysts often want to find out what we believe

is happening in the oil refining and marketing and chemicals segments of the industry. Most analysts know that one of the most significant issues in refining and marketing during the next couple of years will be the supply-and-demand balance as oxygenated fuels enter the marketplace. I also show them supply-and-demand charts for our two key commodity chemicals. These charts position the industry and where we see those commodity chemicals in the industry.

After positioning the company within the industry, I explain some of the good news about the company. I may use colored maps and give the analysts a geography lesson. We will discuss a couple of holes off the North Slope and a few good discoveries.

Investor relations people must be able to handle the good, the bad, and the ugly, however, so I also explain the bad news. The explosion at our Houston chemical complex put us out of the polyethylene-producing business in the United States and destroyed 20 percent of U.S. polyethylene capacity. The television coverage brought our name into focus around the country, but not exactly in the way we prefer to see it. We have to discuss these problems and explain to the marketplace what they cost the company.

The Mock Interview

Smith: *Phillips had excellent market performance during the past five-year period, but most of that occurred during the first three years that the company was working under the discipline of debt. What has happened since then? The performance has gone from flat to down.*

Grigsby: *Originally, we decided to try to grow our way out of the debt problem. We took capital expenditures of about $700 million a year up to a peak last year of about $1.6 billion. The plan was that we would improve the debt-to-equity ratio by improving the equity side of the formula. At the time of the plan, in 1989, we were forecasting oil prices at $25 a barrel and gas around $2 a thousand cubic feet (mcf). Obviously, we were wrong, and that is why the company put a new debt-reduction plan forward. Thinking we had done a good job reducing debt from more than $9 billion down to $4 billion, we stopped the program. We wanted to generate income and thought we could earn our way out of the remaining problem. We were wrong. We missed our estimates, and that is why we are reinstituting the debt-reduction program.*

Smith: *I heard that the North Sea Ekofisk platform is sinking into the ground again. This is a potential area of concern. How bad could this situation become?*

Grigsby: *The Ekofisk reservoir is a chalk reservoir, and as we remove hydrocarbons, the chalk compresses and lowers the ocean floor. The process, called subsidence, continues. The company will present a complete development plan to the Norwegian government by July 1, 1993. In it, we will propose in detail how to address government concerns about safety and the impact of the situation on future production and transportation.*

Smith: *I do not understand what is happening with the preferred issue of Phillips's GPM subsidiary.*

Grigsby: *Phillips originally attempted an initial public offering of GPM common stock in spring 1992. The IPO market had turned down, and gas at that time was $1 an mcf. We determined that the value the marketplace put on the business was not high enough, so we pulled the offering. We are now back in the market with a cumulative perpetual preferred offering of 20 percent of GPM (now called Phillips Gas Company). This approach allows deconsolidating GPM's business for tax, but not for financial, purposes. It is a tax-efficient move to preserve Phillips's tax position.*

Smith: *What are top analysts saying about Phillips?*

Grigsby: *Currently, analysts are in two camps—the pessimists and the optimists. The pessimists say we squeezed all the profits we can from natural gas and we have already been rewarded for that. They say refining and marketing will be flat for another year, petrochemicals will be down for another year, more imports from Iraq are likely, the business will be dog-eat-dog for the next 12–18 months, and there is nothing we can do about it.*

The optimists say more than half of our assets are providing almost no net income today, the stock is trading in the middle $20s even after the negative Norwegian pronouncements about the subsidence issue, and all we have is upside potential. As soon as the refining and chemical businesses come back, they will make a significant contribution. Gas will be stronger in the next couple of years than was thought likely, and the Iraqi oil will make no difference.

Question and Answer Session

Arthur L. Smith, CFA
Edd Grigsby

Question: How do you deal with an investor relations person who has an agenda that is different from your agenda without being adversarial?

Smith: I do not know that you will ever win that battle, but I would try to steer the direction back to your areas of focus. Most companies want to communicate a certain amount of information. Phillips, for example, may work up a slide show and want to go through the information on each slide. If you only have an hour, ask to take a copy with you so you can use the interview to cover your main points.

Question: How do you determine how much of management's story is optimism and how much is truth?

Smith: It takes experience to identify when a company is too caught up in its own story to have a clear perspective on what is really happening. A feel for the marketplace helps. Analysts should also question what other people in the industry think about a company. Successful companies in the oil business are often bad-mouthed by their competitors. Everyone in the industry wants to know everything about the competition, but they do not want to tell anybody anything. It is important to get enough information to determine whether you have a big winner or another flash in the pan.

Moderator: One aspect that I have always liked about the business is that, by virtue of joint ventures and cross ownerships, analysts can validate input. If the cross-check is negative, the analyst must judge, based on the competitive interplay, whether the negative feedback is valid. Analysts will develop different interpretations. In the upstream segment, for example, analysts with different interpretations will have a range of estimated recoverabilities and earnings for Triton Energy and British Petroleum. You can come to some judgments if you look at the patterns over time.

Question: What are the most common mistakes analysts make in looking at the industry?

Grigsby: I do not know of many. Analysts are usually on top of what is happening in our industry. Occasionally, the market gets away from them based on what is happening in the gas business or in the oil patch, but not very often. From the industry's viewpoint, I do not see error as a significant problem. If we are doing our job right, mistakes should not happen.

Question: How should analysts steer through companies that have one-hit wonders (such as Triton) or a quick bust (such as Plains Resources)?

Smith: I would focus on the long-term competitive advantages of the company. Triton was successful not only because of the Cusiana discovery in Colombia but also because financial pressures caused internal management changes and a refocus on the basic business. Triton is a good example of a company that has been able to play in the international arena and farm out prospects to major oil companies. Farm-outs are a great sanity check for explorationists. Getting a major oil company to part with a dollar is not easy today. Focusing on the long-term competitive advantage will help in analyzing whether small companies have the potential to drill 100 wells rather than just 1 well.

Question: Is it more important to emphasize past performance or future projections?

Grigsby: I expect an analyst to come to a meeting understanding what has already happened; I spend my time on the future—making sure our company's corporate strategy is clearly understood. Emphasizing where we are attempting to go is much more important than explaining the past. This does not mean analysts have to accept our strategy, because they may not agree with our premises. If we do not let analysts know where we are trying to go, however, and if they are dealing only with historical data, they may not value our company properly.

I expect analysts to pick up the statistical data on their own. They must make their own price forecasts for crude oil and gas. If my company produces 200,000-plus barrels of oil a day, however, we make sure that nobody is forecasting us to produce 500,000 barrels a day five years from now.

Smith: Ideally, we would like to find a company that has undervalued assets, a good balance sheet, and a good track record and is able to demonstrate that it will take its cash flows and reinvest them at strong rates of return. Analysts recognize that the historical track record of the oil industry has been abysmal. For a decade, the oil industry has generated – 5 percent returns. Oil companies have to show that new drilling, refineries, or petrochemical plants are good places to reinvest cash flows. Because the past record is awful, analysts have to look to the future and determine what will be different in this cycle.

The Natural Gas Industry—Understanding the Basics

A. Paul Taylor, Jr.
Vice President of Corporate Communications
Anadarko Petroleum Corporation

The natural gas story in the United States has unfolded in three chapters: development, regulation, and transition. Now, despite continuing uncertainty, circumstances have converged to create opportunities unimaginable only a few years ago.

The natural gas industry traces its history back to humble beginnings. Until pipelines were built, natural gas was flared off at the wellhead and considered a waste by-product in the exploration for oil in the United States. The construction of interstate pipelines, begun in the 1920s, was a major achievement considering the limited capital resources and poor technology available at the time. Distribution technology and available supply have always limited the growth of the gas industry. The industry only recently could boast of having a transportation network capable of delivering gas nationwide from border to border and coast to coast.

Since the 1800s, the natural gas story has unfolded in three distinctly unique chapters: development, regulation, and transition. **Exhibit 1** summarizes the history of natural gas regulation and price controls.

Development

Unlike the oil industry, in which vertical integration from the wellhead to the gasoline pump is commonplace, the natural gas industry developed into three separate businesses: production, pipelines, and distribution. From 1954 to 1993, gas producers have been finding the reserves and selling gas to pipelines under long-term contracts at prices set by the federal government. Gas pipelines have grown and prospered by purchasing gas in the field and reselling it (with a transportation fee) to captive utility customers in their exclusive market areas. Local distribution companies, the final link with the ultimate end-user, deliver gas to residential, commercial, and industrial customers.

Gas markets grew dramatically in the 1930s and 1940s for three reasons: Supplies were abundant in markets served by the pipelines; gas enjoyed a tremendous cost advantage over such alternative fuels as coal and oil; and local utilities pushed hard to develop their markets by advocating a host of gas appliances that were efficient, clean, and safe.

The historical relationship between gas buyers and sellers changed in 1983, when large industrial customers began to push federal and state regulatory agencies to make substantial changes. Regulatory change and transition continue in the 1990s, and with them comes some uncertainty about the future of the natural gas industry.

Regulation

In the 1800s, states began regulating certain companies that were "affected with the public interest." State governments controlled companies providing grain mills, drawbridges, canals, warehouses, railroads, and gas to ensure that services were neither undersupplied nor overpriced and were available to all.

Companies with operations in more than one state fell under the jurisdiction of the federal government. After a Federal Trade Commission study in 1935 identified specific evils of the interstate pipelines, Congress passed sweeping regulations governing pipelines in the Public Utility Holding Company Act (1935) and the Natural Gas Act (1938). Since 1938, federal regulators have set terms of service, accounting standards, and transportation tariffs for the interstate pipelines.

In 1954, the Supreme Court, in *Phillips v. Wiscon-*

Exhibit 1. History of Natural Gas Regulation and Price Controls

1935	The Federal Trade Commission compiles report on evils associated with natural gas pipeline industry.
1936	Congress passes Public Utility and Holding Company Act, giving the SEC authority to monitor natural gas industry's financial dealings. Title III of draft bill dropped because of opposition to provisions mandating common carriage status for interstate pipelines.
1938	House Speaker Sam Rayburn (D, Texas) rewrites Title III, and bill becomes the Natural Gas Act; imposes significant controls on the natural gas industry—accounting standards and terms of service for interstate pipeline sales.
1954	Phillips Petroleum attempts to raise its natural gas rates and meets opposition in Wisconsin. Supreme Court decides terms of service and rules that the Federal Power Commission (FPC) must regulate the price of gas at the wellhead if the gas is sold in interstate commerce.
1955–77	FPC tries price-control schemes based on costs of service gas, field rates, and area rates. All are too low.
	Gas shortages begin in 1970. Disparity between natural gas prices received in intrastate and interstate commerce discourages development of gas supplies for interstate market. In winter 1976–77, 22 states declare emergency situations, face severe gas shortages for homes and businesses.
1978	Congress passes Natural Gas Policy Act (NGPA), which creates pricing mechanisms for 28 different categories of natural gas, with monthly ceiling prices, and extends price controls to the intrastate market.
1981	Federal Energy Regulatory Commission (FERC) authorizes producers and pipelines to develop special marketing programs (SMPs) to help keep large industrial plants from switching to cheaper fuels. Spot market sales direct from producer to end-user begin.
	President Ronald Reagan deregulates the price of crude oil. Result is ample oil supplies and falling prices.
1981–83	NGPA incentives to explore for expensive gas supplies spur industry boom, but consumer demand drops significantly. Gas deliverability grows while gas demand shrinks; the "gas bubble" arrives.
1983	District of Columbia Court of Appeals strikes down SMPs, cites discrimination because large-scale industrial end-users receive favorable pricing treatment, but local utilities and residential users do not.
	FERC authorizes limited-term abandonments as interim measure to market gas during a period of low demand.
1984	FERC Order 380 takes variable costs (gas, take-or-pay costs) out of pipeline minimum bills, freeing local utilities to buy third-party spot gas.
1985	Five NGPA price categories deregulated; expected surge in gas prices never occurs.
	FERC issues Order 436, which features open-access transportation. Order is widely challenged.
1986	FERC issues Order 451, enabling producers with old gas contracts to negotiate directly with purchasers to obtain higher compensation up to a lawful maximum price. Failure to agree would terminate contract.
1987	Order 436 remanded to FERC for further consideration. FERC responds with Order 500, which provides take-or-pay credits.
	Two additional NGPA categories deregulated.
	Gas shortages reported during a two-week cold period in southern California. Cycle begins again.
1988	FERC issues Order 490, permitting blanket abandonments, which makes possible the release of gas from problem contracts by mutual agreement between producer and pipeline.
1989	Clean Air Act passed. Industries must reduce emissions and convert to cleaner burning fuels.
	Natural Gas Wellhead Decontrol Act (NGWDA) signed; provides for complete decontrol of gas prices by January 1, 1993.
1991	Price controls on newly spudded wells (as defined in NGWDA) expire.
1992	FERC issues Order 636, which calls for unbundling of transportation, sales, and storage services.
	President George Bush signs National Energy Policy.
1993	All NGPA-regulated gas becomes deregulated, free of wellhead price controls.

Exhibit 2. Key Provisions of Mega-Rule—Orders 636 and 636a

- Mandates unbundling of interstate pipeline sales and transportation.
  - Requires pipeline merchants to provide no-notice firm transportation service as an option to other services.
  - Requires pipelines to provide firm and interruptible transportation services equal in quality for all gas suppliers whether purchased from the pipeline or another seller.
  - Makes storage subject to the open-access transportation regulations.
- Issues blanket sales for resale certificates, enabling interstate pipelines to make unbundled firm and interruptible sales.
- Provides pregranted abandonment authority for sales, interruptible transportation, and firm transportation (of one year or less) upon expiration/termination of a contract.
- Authorizes a new capacity reallocation program enabling shippers to release unwanted firm capacity for sale (through the pipeline) to other customers desiring capacity.
  - Requires upstream pipelines to allow all downstream pipelines to assign firm capacity held on an upstream pipeline to the downstream pipeline's firm shippers.
  - Requires downstream pipelines to allocate upstream firm transportation capacity to any of their firm shippers desiring the capacity.
  - Requires that pipelines install user-friendly electronic bulletin boards (EBBs) and post all reassignments first on the EBB and then contract directly with replacement shipper designated to receive an assignment of capacity.
- Requires the use of the straight fixed/variable rate method for billing firm transportation customers unless another method is agreed upon.
- Prohibits tariff provisions that would inhibit development of market centers.
- Establishes policies for recovery of transition costs incurred by pipelines in the process of complying with the new rule.

sin, et al., decided that the federal government should set price controls on natural gas sold to the pipelines in interstate commerce. With the (mistaken) belief that the United States was running out of gas, Congress expanded the use of gas price controls in the Natural Gas Policy Act (1978) and also prohibited gas use by large industrial customers.

Government regulation of natural gas prices was a dismal failure. Gas price controls set too low from 1954 to 1978 caused profound gas shortages in the interstate pipeline systems. Price controls set too high from 1978 to 1985 hurt gas demand and led to a surplus of available supply (the "gas bubble") that plagued the industry for years. Although federal regulators still publish maximum lawful ceiling prices for some gas supplies, the market has effectively ignored price controls since the mid-1980s.

In 1989, with little fanfare and limited debate, Congress closed the book on gas price controls. The Wellhead Decontrol Act provides that all maximum lawful ceiling prices on natural gas expire January 1, 1993, or sooner if contracts are canceled or renegotiated.

Transition

In 1981, the Federal Energy Regulatory Commission (FERC) allowed pipelines to create special marketing programs (SMPs) that offered discounted gas to industrial customers with dual-fuel capacity. Then in 1983, when large industrial customers switched to buying residual oil priced significantly below gas sold by the pipelines, the gas industry began to change dramatically. These large customers wanted more options in buying gas and argued that market signals were thwarted by the historical sales relationships in the gas industry. The battleground for change has been FERC.

In 1985, an unfavorable court ruling, *Maryland People's Counsel v. FERC*, forced FERC to expand SMPs so that the benefits of discounted gas prices would be available to all classes of gas customer. In October 1985, FERC responded with Order 436, a voluntary program through which a pipeline's customers were given open access to the transportation system in order to ship gas they purchased from other suppliers. Despite misgivings, virtually all pipelines embraced the optional program. In 1987, however, the courts struck down Order 436 because FERC had failed to address adequately the pipelines' take-or-pay problems created by the ruling.

FERC next tried Order 500, which formalized the new procedures for open-access transportation and allowed pipelines to recover half the take-or-pay costs incurred for their customers. In 1989, Order 500 was vacated and remanded by the courts because FERC had overstepped its legal authority in granting abandonment of service.

In 1991, FERC officials began working on a comprehensive set of rules to complete the transition from the old system of pipeline sales to a new, restructured gas industry. In a comprehensive ruling

Exhibit 3. Summary of Appeals of Orders 636 and 636a

- 54 petitions for review of Orders 636 and 636a were filed between August 13 and November 2, 1992:
  - 51 in the U.S. Court of Appeals for the D.C. Circuit (Washington, D.C.)
  - 3 in the U.S. Court of Appeals for the Eleventh Circuit (Atlanta, Ga.)
- Because the petitions for review were filed in two circuits within 10 days of the issuance of the orders, the Judicial Panel on Multidistrict Litigation was required to select randomly one circuit court in which all petitions for review would be consolidated.
  - The panel determined that the Eleventh Circuit was the proper court to entertain the appeals.
  - On October 2, 1992, Exxon Company, U.S.A., filed a motion for change of venue with the Eleventh Circuit, requesting that the D.C. Circuit be given jurisdiction over the appeals.
  - Several parties filed answers to Exxon's motion, and Exxon responded to such answers.
  - Exxon's motion still pending (as of late 1992), and the D.C. Circuit cases not transferred to the Eleventh Circuit.
- Several parties filed statements of issues setting forth the issues to be raised on appeal.
  - The filing deadline for a petitioner's statement of issues is tied to the filing date of the petition for review and the court's initial procedural order.
  - All petitioners were to have filed statements of issues by mid-December 1992.

in 1992, Order 636 required pipelines to unbundle their various services (gas gathering, compression and transmission, storage, balancing, etc.) and price each service separately to customers, who could then rebundle the services to suit their particular needs. **Exhibit 2** summarizes the key provisions of Orders 636 and 636a (called the Mega-Rule).

The Gas Industry of the Future

After almost 10 years of regulatory transition and uncertainty, the natural gas industry is poised for greatness in the 1990s. Low prices and expanded competition in the 1980s helped push gas demand up by 20 percent between 1986 and 1991, despite two recent mild winter seasons. Further demand growth in the 1990s is probable under the terms of the Clean Air Act Amendments of 1989. Demand for natural gas, the cleanest burning fossil fuel, should increase as industrial plants move to comply with the act. Efforts to reduce U.S. dependence on foreign oil supplies should also result in expanded use of natural gas.

Much uncertainty remains, however. Some companies, including local utilities that object to the rate methodology FERC prescribes in the order, are challenging Order 636 in the courts. **Exhibit 3** provides a brief summary of the court challenges.

Regulators must adopt rules that satisfy customer needs for flexible gas sales and transportation services, that pass court review, and at the same time, that maintain the financial viability of the pipelines. Natural gas prices must move higher and stabilize if gas producers are to have the financial incentive to develop the U.S. gas resource base and to ensure that gas supplies are adequate to meet rising demand. Senior managers in the gas industry's three sectors must quickly come to grips with the industry's fundamental problems, or the opportunity presented to the gas industry will be lost.

Factors Affecting the Natural Gas Industry

David N. Fleischer, CFA[1]
First Vice President
Prudential Securities, Inc.

Portions of the natural gas industry are poised to take advantage of key forces that drive revenues and profits. Although the utility mentality of the past has not totally departed, the opportunity nevertheless is great today for those companies that reduce costs, increase efficiency, use technology wisely, and provide a high level of customer service.

The natural gas industry has faced numerous challenges in recent years, and companies have handled the challenges in different ways, with drastically different results. Thousands of exploration companies went out of business during the 1980s, when natural gas prices fell and the fundamentals deteriorated. Other companies—Burlington Resources, for example—used new technologies, cut costs, and grew by filling the vacuum left by departing competitors that were unable to cope with the industry's difficulties.

Some companies, such as Columbia Gas System and Transco Energy, mismanaged their regulatory relationships and paid a high price in the form of take-or-pay and other costs. Such companies proved the maxim that "more money is made in Washington than in the marketplace." In contrast, Enron and Coastal consistently earned returns far above the allowed rates of return while avoiding the out-of-pocket costs that buried other companies. An understanding of the Washington factor—and the many others—will allow analysts to evaluate the importance of factors affecting the industry and use available information on individual companies to evaluate their chances for success in the future.

Perception versus Reality

Bear in mind that, in the natural gas industry, reality is what it is perceived to be—not necessarily what an all-knowing observer would determine it to be. In a period of just six months in 1992, for example, supply in the gas market went from being perceived as a surplus to being perceived as in balance to being perceived as a shortage. In reality, the surplus, if it existed, was small, but the perception of surplus, accentuated by warm weather and a herd mentality, ballooned. The perception of supply shortages in late 1992 was probably equally overblown.

Perceptions and misperceptions can influence the actions of government, regulators, companies, and customers. During a period of perceived natural gas shortage from 1975 to 1978, coincidental with three cold winters in a row, most observers believed the country was running out of natural gas. In reality, the gas industry had run out of incentive to look for natural gas. To protect consumers, the government had kept gas prices artificially low and, as a result (see **Figure 1**), reserves had fallen steadily since the early 1970s. Exploration for natural gas was so limited that the available supply in those years was not enough to meet the demand of all customers. Industrial consumption of natural gas, shown in **Figure 2**, began a decade-long decline that, together with reduced utility consumption, accounted for most of the decrease in U.S. gas consumption.

In 1978, the government passed the Industrial and Powerplant Fuel Use Act, which allowed newly found gas to be sold at prices sharply higher than old-gas prices while continuing to control prices of previously discovered supplies. As shown in **Figure 3,** the price of natural gas rose from about 80 cents a thousand cubic feet (mcf) in 1978 to $2.50 in 1982. The market responded with a decrease in demand, particularly in the industrial and utility segments, that resulted in a dramatic increase in the supply of new, but very expensive, natural gas.

The country was thus saddled with a surplus of high-priced natural gas that became known as the "gas bubble." This bubble was to be a temporary

[1]Mr. Fleischer now is a vice president at Goldman, Sachs & Company.

Figure 1. Summary of Annual Estimates of Proven Reserves of U.S. Natural Gas, 1960–91

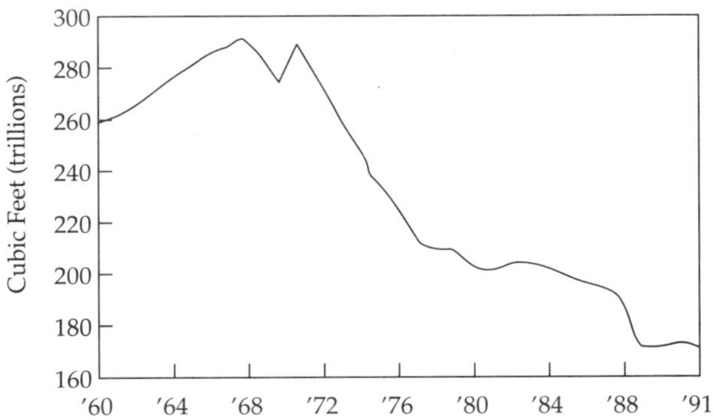

Sources: American Gas Association, American Petroleum Institute, and Canadian Petroleum Association for 1960–75; Energy Information Administration for 1976–91.

Figure 2. U.S. Industrial and Residential Natural Gas Consumption, 1960–92[a]

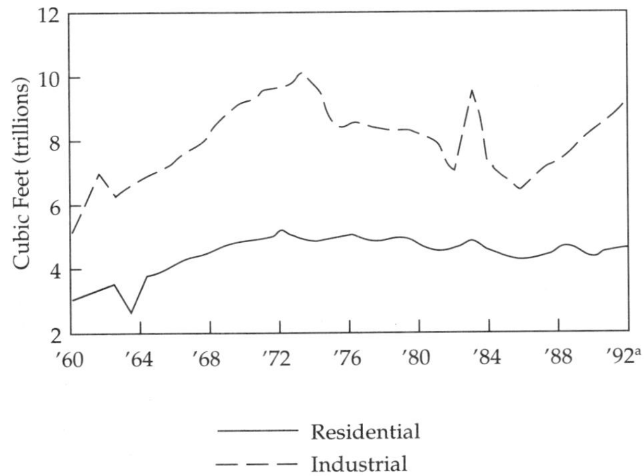

Sources: Energy Information Administration; author estimates.
[a]Estimate.

respite from inevitable gas shortages, but the surplus lasted for more than a decade. In 1985, the government acted to decontrol old natural gas supplies and unleash the power of the free market. Prices, instead of increasing as many had predicted, fell, and they kept falling as the industry reduced costs and became more competitive.

The natural gas industry also contributed to the demand problem during the 1980s. By talking about the need for higher prices to stimulate the search for more supplies for the long term, the industry chased away present and potential customers, who became convinced that natural gas was an expensive and unreliable fuel. Eventually, the free market began to win over misplaced rhetoric and misplaced government action. Lower costs and prices and adequate supply began to win back customers.

One factor has been working in favor of stability in natural gas prices: The New York Mercantile Exchange (NYMEX) brings buyers and sellers of gas together through daily trading of natural gas futures contracts. NYMEX removes some of the pricing inefficiencies that have existed in the natural gas market and is the source of many market signals.

Many natural gas stocks suffered because of the misperceptions of the 1970s and 1980s and the stock market responses to them. Perceptions had displaced reality and influenced actions and prices accordingly.

Fruitful investing in this industry depends on knowing both the real status of the key determinants of industry success and knowing the perceptions about those determinants. To profit from a changing cycle, investors must be early but not too early. They need to know the point at which a perception is changing from, for example, one of a surplus to a shortage, of unfavorable regulations to more favorable regulations, or of mistake-prone management that has a "utility mentality" (an attitude that pipelines simply buy gas from producers and pass the price through to consumers) to a perception of savvy, competitive management. In each of these important aspects of the natural gas industry, progress toward a healthier industry has been steady and impressive during the past several years, but investor perceptions and valuations do not change until the shift in the cycle becomes obvious, and then they change suddenly.

Factors Affecting the Gas Industry

Until recently, external factors have not been especially favorable for the gas industry. Fortunately,

Figure 3. Average U.S. Natural Gas Prices, 1973–92[a]

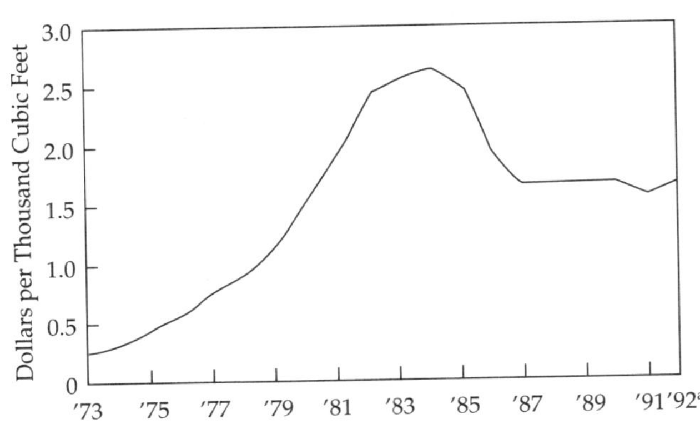

Sources: Energy Information Administration; author estimates.
[a]Estimate.

many of these factors have turned more positive in recent months.

Long-Term Structural Trends

Long-term trends in economic growth, competition (particularly with respect to alternative fuels), and the growth of industrial and utility demand are primary influences on future demand for natural gas in the United States.

■ *Economic growth.* Energy consumption in the United States, as shown in **Figure 4**, has historically tracked GDP growth. In the past, when energy prices were extremely low, energy consumption grew faster than GDP growth; recently, with high energy prices and the drive toward conservation, growth in energy consumption has slowed markedly. Between 1973 and 1983, U.S. energy consumption actually fell, although GDP rose by about 20 percent. Some growth then resumed, because the easy gains in conservation had been made. In the long term, however, significant growth in energy consumption cannot be assumed in a low-growth, service-oriented economy.

World economic activity has a modest impact on the domestic natural gas industry. If economic growth leads to higher oil demand and increased oil prices, then natural gas prices can rise during the spring and summer months, when gas must be price competitive with fuel oil. During the rest of the year, gas is primarily a heating fuel that sells at a price that depends on the domestic supply-and-demand balance and with no relationship to either the price of oil or worldwide economic activity.

■ *Competitive forces.* In international competition, Canada has gained considerable market share in the U.S. natural gas market, primarily because it has been a more reliable supplier of natural gas than many U.S. suppliers. Canada's share of the U.S. market grew from 4.6 percent in 1984 to roughly 10 percent (or slightly more) at the end of 1992.

Mexico, which has substantial energy reserves, particularly in the south, and once exported natural gas to the United States, is now an importer of U.S. gas. Its border areas are likely to become even larger customers in the future.

The natural gas industry lost market share to oil and other fuels between 1973 and 1986. As **Figure 5** indicates, in 1973, the natural gas industry accounted for about 31 percent of total U.S. energy consumption; by 1986, its market share had fallen to about 22 percent. Since the 1986 repeal of the Industrial and Powerplant Fuel Use Act of 1978, industrial and utility demand for gas has been rising; overall natural gas consumption has been rising about 3.5 percent a year and now accounts for almost 25 percent of total U.S. energy consumed.

President Bill Clinton is believed to favor tax credits and other incentives to encourage natural gas vehicles and using natural gas for commercial air conditioning in order to reduce foreign oil imports. Given the huge changes that would be required to add compressed natural gas stations to the infrastructure and to retool U.S. automobiles, this program is unlikely to create significant demand within the next four or five years. Investors are likely to react positively to the announcement of such a program, however, because it would indicate confidence

Figure 4. Growth in U.S. Energy Consumption versus Real GDP Growth, 1974–92[a]

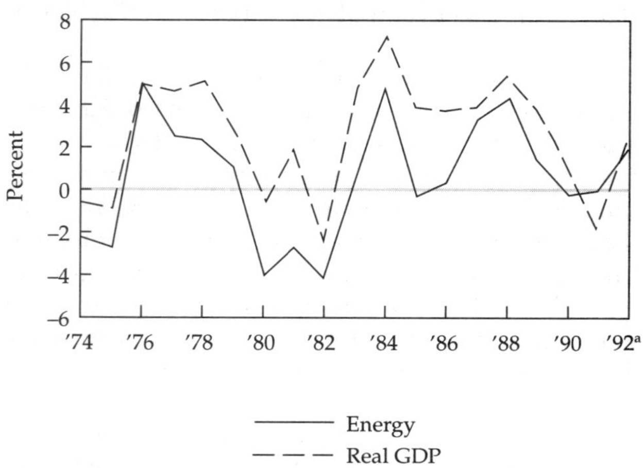

Sources: Energy Information Administration; author estimates.
[a]Estimate.

Figure 5. Natural Gas as a Percent of Total Energy Consumption, 1973–92[a]

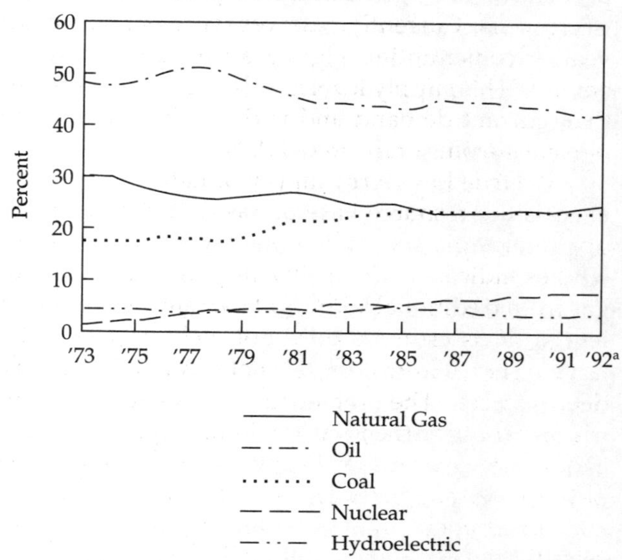

Sources: Energy Information Administration; author estimates.
[a]Estimate.

in the future of the natural gas industry.

Importing liquefied natural gas (LNG), once thought likely to become a big business in the United States, will probably be relatively unimportant. Although four major import terminals were constructed in the 1970s (two are currently not operating), import levels remain small because of the high cost of importing—$2 an mcf of out-of-pocket costs, assuming all terminal costs are sunk costs. Because of the substantial cost reductions the domestic gas industry has made, LNG imports are primarily used to meet peak winter demand.

- *Industrial and utility demand.* The industrial and utility segments of the natural gas market appear to be the only segments that will show significant growth in demand during the next five years. Cogeneration and independent power projects represent the major growth areas; the growth in demand for electricity and the retirement rate of nuclear power plants will determine the potential for natural gas. The key determinant in winning certain major projects rests on the industry's ability to change the perception of natural gas as an unreliable fuel. The coal industry is willing and able to offer 30-year contracts, and utilities can keep large inventories of coal on site for security of supply, something the natural gas industry cannot offer. The natural gas industry must convince its customers, therefore, that reasonably priced supplies are and will be available well into the future. In short, a new long-term market must be developed.

Supply

Investors tend to think of supply in terms of proven reserves, gas that can be produced from existing wells. Currently, reserves represent only eight years of consumption (eight years of current deliverability). This supply level is only 5–8 percent greater than current demand and is down from the 25–30 percent surplus of the mid-1980s.

The true long-term supply of natural gas, however, is the resource base of gas that can be developed, and this supply is potentially huge. Recent studies indicate a 65- to 80-year potential supply of gas from traditional sources at current consumption levels. This estimate does not include potentially large reserves from coal, oil shale, salt brine, or very deep sources. The problem is that, because customers buy most of their gas on 30-day spot deals, gas producers see no reason to develop the resource base beyond currently needed supply. Customers are concerned about the availability of long-term supplies (in light of the industry's history as an unreliable supplier) and the price of future supplies (because obtaining long-term price guarantees is impossible in most cases).

The current level of deliverability is believed to be essentially in balance with current demand. Nevertheless, external factors can shift the perception of shortage or surplus. A cold winter can easily change the perception toward one of shortage and cause gas prices to spike to more than $3 an mcf. This reaction would not change the fundamental supply-and-demand balance, but it would create a selling opportunity for investors.

Government Actions

Through legislation and regulation, governments are a major influence on the natural gas industry, and because the political environment in the United States has been relatively hostile to the oil and gas industries, the major companies have had significant incentives to divest U.S. oil and gas assets. The proceeds are then moved overseas to countries that offer large acreage concessions in less mature basins, tax incentives, and royalty holidays.

Among the disincentives to oil production in the United States are the following:

- *Significant acreage is off-limits to drilling.* In such areas as offshore California, major discoveries prove difficult to bring into production.
- *Stringent environmental rules can increase costs and lead to distorted signals about the true supply and demand.* Multiyear delays in obtaining environmental approval to build pipelines to the gas-short northeastern United States, for example, led ultimately to excess costs and surplus capacity. Almost in unison, the natural gas companies recognized the need for increased capacity in the Northeast, responded with proposals, and began a long wait for approvals of their environmental impact statements. Each company was committed to its project, regardless of whether the projects made good economic sense. The result was eventual overcapacity in the Northeast.
- *U.S. tax policy has been less favorable than policies abroad.* The alternative minimum tax has been onerous to the oil and gas industries and has reduced the capital available for reinvestment. Ordinary drilling costs are a tax preference item. Tax burdens have been high on domestic exploration companies during the past decade, even though earnings have been virtually nonexistent because of the difficult environment. In contrast, foreign countries have been willing to use tax policies to encourage investment.
- *Government rules and regulations have been capricious and uncertain.* For example, with a stroke of the pen in 1984, the Federal Energy Regulatory Commission (FERC) created a situation in which pipe-

lines that had entered into long-term supply arrangements with producers found that their customers were no longer obligated to buy the gas. Huge financial obligations followed. The Clean Air Act was supposed to give the natural gas industry a chance to compete with coal for certain utility loads, but to protect jobs in local mines, 11 coal-mining states have passed laws requiring installation of scrubbers on local coal plants, thus removing utilities' least-cost natural gas option: If they have to install scrubbers anyway, utilities will continue burning coal.

Regulation has radically improved with the issuance of Order 636. Also known as the "restructuring rule," the order will change the way the natural gas industry is regulated and the way it functions. Essentially, many functions are being deregulated. More importantly, after about eight years without a well-defined regulatory road map, the industry will at least have some sort of rule book on how to play the game.

Most investors do not perceive the current regulation to be much of an improvement, however. They are apparently in a wait-and-see mode and will believe the change is for the better only after they see the results. If the new regulation improves industry fundamentals, investors will have a valuable opportunity in natural gas stocks as price-to-earnings multiples are adjusted. The consequences of Order 636 will be different, however, for the different segments of the industry.

Natural gas pipelines will be the big winners. The regulatory changes will balance the rules and level the uneven playing field these companies have had to deal with recently.

Exploration companies and major oil companies that produce large quantities of natural gas will also win in the new regulatory environment. They will regain marketing control over their own reserves and, by choosing what gas to produce and how, capture additional marketing profits.

Local distribution companies (LDCs) will apparently be the big losers in the new regulatory environment. In the past, the pipeline companies had an obligation to serve the LDCs even though the LDCs had no obligation to buy from the pipelines. In the future, LDCs will have to provide their own gas supply. Distribution companies will have every opportunity to make the same mistakes pipelines made in supply management—paying too much and having too much at various times to avoid being short of gas at other times. Public utility commissions, using 20/20 hindsight, will find various costs imprudent and force the LDCs to absorb many of them.

Weather

If global warming is occurring, demand for natural gas heat may fall gradually over the years. As with many other external factors, this development need not, however, be a negative for the industry. It could, now that cost-efficient commercial natural gas air-conditioning equipment is available, be an opportunity to spread natural gas consumption over the course of the year by aggressively pursuing air-conditioning load.

On the other hand, the eruption of Mount Pinatubo in June 1991 may be having a cooling effect on weather patterns; apparently, the sulphur dioxide particles produced by volcanic eruptions reflect sunlight back into space. Two major eruptions in the 19th century, Krakatau and Mount Tambora, were followed by bitterly cold winters. In 1992, every month from February through November was colder than normal in the Northern Hemisphere.

Even if the weather is not particularly cold, the stock market and investors are likely to perceive any cold period that occurs—particularly early in the heating season—as an opportunity to boost prices. Therefore, perceptions about the possibility of cold weather may be as important as the reality.

Availability and Cost of Capital

After a difficult decade of weak earnings and write-offs from regulatory burdens and misguided diversification, the natural gas industry is weighed down by high debt levels and reduced credit ratings. This condition translates into decreased and costly access to the credit markets at the exact time when the industry should be investing to develop the resource base. More equity capital is needed, but higher profits, less profit volatility, and more confidence in the future are prerequisites to investors providing that capital.

The result could be a time lag between rising demand for natural gas and obtaining the capital necessary to meet the demand. The depressed rig count shown in **Figure 6** may remain so for some time. The major oil companies are continuing to divest in the U.S. gas industry, and the small exploration companies are virtually excluded from the capital markets. The rich companies may get richer while the small and leveraged companies remain capital constrained. For now, investors would be well advised to stick with the better capitalized companies, which have the most flexibility.

Interest rates help determine which sources of natural gas will be developed. High discount rates make the time value of money an important factor and thus increase the importance of recouping an investment quickly. The high interest rates of the

Figure 6. U.S. Rotary Rigs in Operation, 1973–92[a]

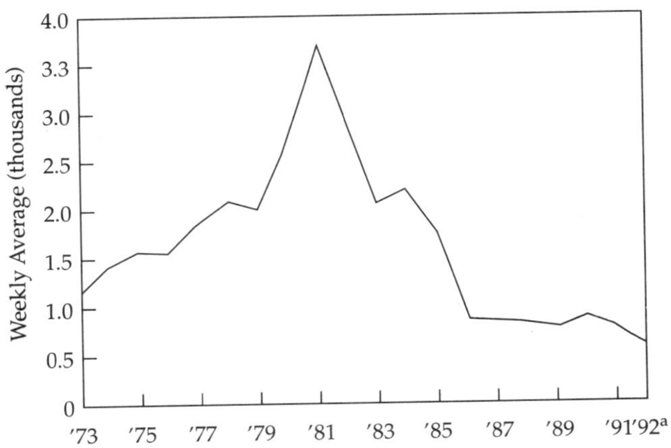

Sources: Energy Information Administration; author estimates.
[a]Estimate.

early 1980s provided an incentive to invest in the high-deliverability fields of the Gulf of Mexico, for example, even though finding costs were higher there than on shore. In today's environment of lower interest rates, reserves that might take 20 years to extract are more economical than some high-deliverability fields if the finding cost of the longer lived reserves is low.

The costs of equipment, such as rigs for drilling wells, can rise and fall a great deal. During the past decade, costs have fallen in many categories; suppliers to the natural gas industry have had to monitor costs closely to stay in business. In the future, equipment costs are likely to rise somewhat as demand increases and less efficient rigs and crews are employed.

Responding to the Challenges

Although external factors are outside a company's control, managers do have some power in some situations over how these factors influence company fundamentals. For example, managers can structure rate designs to be more or less indifferent to weather, and difficult regulatory orders can be turned into positives. Only a poor management repeatedly blames the weather, excess supply, or bad regulations for earnings problems.

Portions of the natural gas industry are farther along than many investors realize in benefiting from key driving forces that will add to revenues and profits. Unfortunately, the headlines have focused on companies that have blundered, but a new mentality exists at certain companies as managers approach business on a free-market basis. That is not to say that the utility mentality of the past has totally departed. Nevertheless, the opportunity is great today for those aggressive companies that separate themselves from the herd by reducing costs, increasing efficiency, using technology better than in the past, and providing a high level of customer service.

Asset Allocation

The most basic management decision concerns which asset classes merit investment and in what proportion. Asset allocation should be the topic of the first question analysts ask when interviewing the managers of natural gas companies. The traditional choices for natural gas companies have been among exploration and production, gas gathering and processing, transmission, oil services, and distribution. Marketing is the new and increasingly important area of concentration, but staying within a company's areas of expertise has proved to be important.

For example, companies such as Enron have hedged the risks in their commodity gas production by investing in natural gas liquids, where profits often move in the opposite direction of natural gas prices because natural gas is a major part of the cost structure of LNG operations.

The excessive use of leverage has been a pitfall for many managements, who tried to pick the bottom in the energy markets but then found they had overleveraged their companies with premature and costly investments. No one can call a market top or bottom, so betting the company is never a good idea.

Diversification has been a disaster for most natural gas companies that pursued it. Investors should be wary of managers who say they want to pay top dollar to get into a business they know little about to "diversify shareholders' investments." What the managers mean is that they do not know how to invest money in the businesses they know best. Pacific Enterprises, for example, invested $850 million to diversify into retailing assets and then sold them for about $50 million after suffering losses and write-offs. The stock, which was $55 a share at the time of the bid, was $19 a share in November 1992, even though utility earnings grew steadily during that time.

The Williams Companies' diversification into telecommunications is a more successful example of diversification, although the share price at the end of 1992 indicates that investors do not place much value on Williams's telecommunications subsidiary. Williams built the new subsidiary from the ground up with a concept that no one else had used: running fiber-optic cables through abandoned pipelines on existing rights-of-way. As a result, Williams built its network at a substantially lower cost than its competitors. It later acquired telecommunications equipment and technology companies at well below replacement costs. Because of its low-cost system, Williams has built-in exit strategies; even if it is not completely successful, investors will likely come out ahead.

Cost Structure and Efficiency

Low-cost producers usually survive the difficult periods and excel in the better times, even though more leveraged companies may appeal to investors trying to take advantage of a change in fundamentals. Being low cost is not a result of simply cutting employees or other costs indiscriminately. In this capital-intensive business, using the right amount of technology, making the right asset allocation decisions within segments, and providing leadership for workers to accomplish all the small items add up to a competitive advantage. In addition, although warm weather reduces natural gas demand and prices, companies can offset this risk by selling gas on longer term, fixed-price contracts or creating rate structures that are relatively indifferent to volumes.

Natural gas finding costs vary dramatically among companies, even though the highest cost companies have been steadily going out of business over the past decade. (Of some 14,000 exploration companies that existed a decade ago, only 4,000 remain.) The best companies consistently find gas at 50–70 cents an mcf. Many others have finding costs of $1.50 or more. The difference lies in the management.

Perhaps the best example of a successful exploration company's turnaround is that of Enron, a company that was decidedly average in 1988 before the arrival of new management. By using technology judiciously and taking advantage of the departure of major oil companies, Enron has shown that a company can win, even in periods of falling prices and weak markets, by becoming substantially more efficient than its competitors.

Companies can also use acquisitions to achieve efficiencies, but acquisitions do not always create value. Both Transco and Panhandle Eastern, for example, acquired pipeline companies to increase the throughput of their pipelines. The concept—that of pairing an underutilized midwestern pipeline with a northeastern pipeline to growing markets—was sound, but Transco and Panhandle Eastern paid top dollar to acquire pipeline systems that could earn only a regulated return, which is questionable. In both cases, the stocks fell considerably when expectations were not met, and the managers who made these investment decisions are no longer running the companies.

Technology

Technology has contributed greatly to cost reduction in the natural gas industry. The use of two- and three-dimensional seismic, bright spot, and other technologies has resulted in far fewer dry holes than in the past, and more oil and gas is being found per foot drilled. In addition, such technological improvements as horizontal drilling and enhanced well-completion techniques have improved the economics of searching for and producing natural gas. Companies in the industry use these technologies differently; the best companies use technology judiciously, by purchasing only the technologies they need and making cost-efficient choices.

Marketing and Customer Service

Although somewhat different concepts, marketing and customer service fit together. The best way to sell to customers is to know their needs, serve them well, and bring credibility to the relationship—in summary, to provide a high level of customer service.

Credibility has been in short supply in most segments of the natural gas industry for a long time. First, natural gas ran out of supply for some customers; then, prices shot up to astronomical levels, pipelines reneged on take-or-pay contracts, contractual relationships changed as regulation continued to change, and meanwhile, companies talked about the likelihood of declining supplies and higher prices. This situation did not promote good customer relationships.

If gas is to win more business, mechanisms must be created to supply both price and volume guarantees that are credible to customers. Both sides would like to get away from transacting natural gas sales using 30-day spot prices in the deals, which are frequently on a best-effort basis.

Although proving reliability will take time, some companies are already gaining credibility within their own spheres. The best example, again, is Enron, which has acquired a long list of admirers among its customers. The result for Enron has been substantial gain in market share and growth in profit.

To provide the requisite level of customer service, the industry needs to shed its utility mentality. After years of operating as regulated utilities, producers now control what they sell and how they sell it. Pipeline companies will still be regulated transporters of gas, but many other functions that were regulated will now be free-market activities providing every opportunity to win or lose. In the past, the gas industry did little to market its advantages as a clean-burning, readily available, and low-cost fuel. Marketing may now become a key function.

Conclusion

Natural gas companies have an unprecedented opportunity to differentiate themselves in an environment in which the less efficient and less competitive companies will be the losers and the best competitors can make significant profits. After a difficult decade of declining prices, uncertain volumes in a surplus environment, and huge regulatory and other burdens, the industry is apparently about to emerge from its malaise. For investors, opportunities frequently appear when an inflection point arrives in a cyclical industry. Such opportunities await today's investor in natural gas.

Question and Answer Session

David N. Fleischer, CFA

Question: On the basis of Btu (British thermal unit) equivalency, how does the price of natural gas compare with that of coal?

Fleischer: Given the various prices of natural gas and coal, which depend on the various markets and types of contracts, generalizing is almost impossible, but having said that, I'll venture that coal at the mine is generally lower cost than natural gas. On an all-in cost basis, however, natural gas is cheaper, because small natural gas cogeneration units can be built quickly, and the capital costs involved are lower than for coal-fired facilities. The environmental expenses of burning coal are high. At less than $3 an mcf for natural gas at the wellhead, gas can be competitive with coal in most uses, but the cost of transporting gas for long distances is higher than the cost of using local coal. Thus, location makes a difference. With gas, the environmental costs and the costs of construction delays are minimal or nonexistent.

Question: How much natural gas demand can be switched to fuel oil, if prices so dictate?

Fleischer: There are as many different answers as there are studies on this. Half of industrial and utility demand and a few commercial customers are technically capable of switching, but the real question is, "Who can convert on a dime?" With high sunk costs, only a small number of customers can switch immediately if the price of natural gas shoots up.

Industrial and utility demand account for nearly half the total natural gas consumption in the United States. On an annual basis, 1.5 trillion cubic feet of demand would likely be lost if gas prices moved up sharply. Big utilities such as San Diego Gas and Electric will shut down a gas burner and bring on fuel oil, but most prefer to act on the basis of longer term trends. They do not want to switch for a day or a week. In the long run, however, if gas were to become very expensive, half of industrial utility demand would disappear.

Question: What are your views, with respect to the availability of capital, regarding companies that use production payments to acquire reserves?

Fleischer: Not many companies are using this strategy, but it is a great way to take advantage of the stronger marketplace and to funnel capital to producers. Enron created a unique concept that moves capital to the industry. Enron takes advantage of its healthy balance sheet to obtain capital cheaper than exploration companies can—if they can get it at all; the company writes strong contracts and receives strong guarantees. Enron requires that the gas producers provide double the required reserves to back what it guarantees to customers. The money comes through a separate financial subsidiary, and Enron funnels it to the producer, which then has more money for drilling. In return, Enron acquires reserves at known and reasonable cost and locks in a significant margin. In many cases, the producer could not have financed the transaction itself at all.

Production payments are an effective way to funnel more capital into the industry, and we will likely see many more of them in the future.

Question: How do you respond to the theory that the gas bubble burst in the winter of 1987, when the industry began to have shortages on peak days?

Fleischer: The real questions are, "What is a surplus?" and "What is a shortage?" The industry has both a surplus and a shortage at all times. Producers think there is a surplus if they cannot sell all the gas they want on most days of the year. Customers think there is a shortage if they cannot get all the gas they want every day of the year. An industry that can meet demand for the one peak day or week in the winter is inefficient. The gas industry designed the system of interruptible customers to deal with this situation: Industrial and utility customers pay lower rates for the privilege of being thrown off the system in the winter. A healthier system would be to switch these interruptible customers to other fuels during peak months and serve them only during the rest of the year.

The surplus-versus-shortage question goes back to perception. On an annual basis, the surplus is probably only 6–8 percent. Today's level does not appear to be a surplus at all, but a normal working supply.

Question: Please explain the competitive pitfalls and opportunities facing the pipeline industry under Order 636. In what respects might investors be surprised at what evolves in an unregulated environment?

Fleischer: Order 636 is a complicated document. Every aspect of the order discusses flexibility and allowing the customers and the pipelines to reach agreement on how to function in this brave new world. Of course, whether any regulatory order succeeds in accomplishing what it is supposed to accomplish is the result of two factors: how it is written (636 is written with a fair degree of flexibility) and how it is implemented/interpreted. Most pipelines have submitted their documents on how to adopt Order 636, and they have many different approaches. We will probably be favorably surprised at the positive and imaginative solutions that companies develop to earn more money in the future. We will be disappointed with others who are not so clever about how they structure their businesses. That is why there will still be some big differences, opportunities, and risks in the business.

Question: Under Order 636 and straight fixed/variable (SFV) rates, is the notion of gravitation toward the mean—the weak get stronger and the strong get weaker—correct or overblown? Also, do you agree that gas brokers will not be the business of the future?

Fleischer: Independent brokers would have difficulty gaining the credibility of the bigger players. There will be successes but probably very few. Order 636 creates a tremendous opportunity for pipeline companies that have the necessary market knowledge to be successful on the marketing side.

As far as who wins and loses, not everyone will gravitate toward the middle; there will be as many differences in winners as before. The SFV rate design and other attributes of Order 636 may keep the risk down for certain players, and the freedom the order creates will create an abundance of opportunities—particularly for the smartest managers. The filings with FERC to adopt Order 636 contain some innovative sources of earnings that will eventually become apparent. In short, the opportunities exist, but investor perceptions are still keying on the risks.

Question: Will the bigger companies be better off under Order 636?

Fleischer: Not necessarily. I see opportunities for both the smart small companies and for larger companies to excel.

Question: Do pipeline companies have more event risk under Order 636 than previously?

Fleischer: The risks to pipelines under 636 are small. Investors are concerned about transition costs, but a nearly 100 percent passthrough of such costs to customers is likely. The other provisions and possibilities for pipelines also appear to be positive.

Question: Will pipelines have problems collecting transition costs paid to producers?

Fleischer: Order 636 allows for 100 percent passthrough of transition costs, and I believe that well over 90 percent of such costs will actually be collected.

Question: Will President Clinton change Order 636? If so, what will change?

Fleischer: An activist new administration could theoretically do anything. Order 636 has been years in the making, however, and is, in my judgment, a fair and well-reasoned document. I expect it to be fully implemented in 1993 and with only modest changes.

Question: Do you expect the pipeline industry to consolidate?

Fleischer: I expect more strategic alliances and coordination, but the days of paying up to buy rate base are probably behind us.

Question: What effect will the North American Free Trade Agreement (NAFTA) have on Canadian and Mexican gas imports and exports?

Fleischer: I doubt that NAFTA will have any significant impact on the flows of natural gas. Canada and the United States are already functioning in a virtual free market. The opening of the Mexican marketplace is likely to lead to more U.S. gas flowing to Mexico.

Question: What is the next major problem facing pipeline companies?

Fleischer: How to spend all the profits they will soon be making! Now that the major problems of the past are almost solved, the most important challenges involve future investment decisions, asset allocation, and management of growth in an improved marketplace.

Interpreting the Natural Gas Numbers

John E. Olson, CFA
First Vice President of Securities Research
Merrill Lynch & Company, Inc.

In the 1980s, massive regulatory and legislative changes caused the convulsions that resulted in the natural gas industry's current cycle. These included changes to industry structure as well as to the rate-making process.

To comprehend the state of the natural gas business, analysts must understand rate-base economics and rate-making trends. The rate base is the net investment in pipelines, storage, or assets of a local distribution company (LDC) that can provide a regulated return to utility investors. The gas industry has about $150 billion invested in natural gas reserves in the ground. These reserves are unregulated. Another $100 billion of regulated pipeline assets translates, after depreciation, into an approximately $40 billion rate base. LDC assets amount to another $50–$75 billion of rate base. All fixed costs, profits, and related taxes compose the cost of service and are recovered out of pipeline or LDC tariffs. The pipeline and distribution industries together currently invest about $10 billion yearly, a record level in the industry's typical 10-year economic cycle; depreciation charges are $3–$4 billion annually.

Analysts often categorize the natural gas markets as upstream, midstream, downstream, or endstream. Upstream refers primarily to exploration and production but also includes gathering, natural gas liquids (NGL), gas brokering, and futures hedging. Midstream refers to pipeline companies with some ancillary businesses attached to them. Downstream refers primarily to local distribution, and endstream is cogeneration and independent power. Companies strive to integrate the various businesses and collect the value added all the way down the chain. The asset combinations that are occurring or are on the drawing board reflect the drive to capture these integration opportunities.

As **Table 1** demonstrates, the midstream—pipelining—generates about 69 percent of total operating income; the remainder comes from exploration and production, local distribution, refining and marketing, and oil services. The pipeline companies are actually quite diversified, however, and the table includes their earnings from more than pipelining. If Tenneco, the most diversified pipeline company, were included in the calculation, for example, pipelines would provide only about 53 percent of total earnings. The producer/gatherers are the most diversified of the natural gas companies and earn more from local distribution than from exploration and production. **Table 2** reveals a similarly wide distribution of the assets among these industry segments.

Economic cycles in this industry last about a decade. The industry fared well in the 1950s, suffered in the 1960s, and then recovered in the 1970s. During the era of partial deregulation after 1984, natural gas was not a profitable place to invest. In 1992, however, supply and demand began to converge, and the outlook has thus become much more positive.

This presentation focuses on the role of rate making in the natural gas industry and on the regulatory and legislative changes in rate making that have created the current cycle. During the 1980s, the Federal Energy Regulatory Commission (FERC) made fundamental changes in the rules that had governed the industry for the previous 45 years. Neither managers nor investors fully understood all the convulsions that would ensue, because no precedents existed. Everyone learned the hard way. For better or for worse, these convulsions have created the newest industry cycle.

Structural Changes

Before 1985, natural gas flowed through highly structured channels, mostly under a wide umbrella of federal regulation. It was a merchant business, in which a pipeline company bought gas from a pro-

Table 1. Distribution of Operating Income by Activity—Major Companies by Industry Segment, 1991

Industry Segment and Company	Net Operating Income ($millions)[a]	Percent of Net Operating Income: Pipeline	Exploration and Production	Local Distribution	Refining and Marketing	Other
Producers/gatherers/NGLs						
American Oil & Gas Corp.	$ 30	96%	—	—	—	4%
ENSERCH Corp.	133	—	24%	87%	—	(11)
Equitable Resources	117	—	40	60	—	—
Mitchell Energy & Development Corp.	116	78	36	—	—	(13)
Questar Corp.	130	33	26	42	—	—
Southwestern Energy Co.	42	—	67	33	—	(1)
Total and percent of total	$ 567	28	32	45	0%	(5)
Pipeline owners/operators						
Arkla	228	26	9	70	—	(5)
Coastal Corp.	545	82	8	—	5	6
Columbia Gas System	233	55	(2)	49	—	(2)
Consolidated Natural Gas Co.	230	38	9	48	—	5
El Paso Natural Gas	180	100	—	—	—	—
Enron Corp.	498	66	13	—	—	21
Panhandle Eastern Corp.	455	86	—	—	—	14
Sonat	203	69	22	—	—	9
Transco Energy Co.	149	63	10	—	—	26
Williams Companies	277	81	—	—	—	19
Total and percent of total	$2,997	69%	7%	13%	1%	10%
Large LDCs						
Atlanta Gas Light Co.	121	—	—	100	—	—
Brooklyn Union Gas Co.	122	—	—	100	—	—
MCN Corp. Holding Co.	99	—	—	94	—	6
National Fuel Gas Co.	122	34	9	58	—	(1)
NICOR	190	—	2	92	—	6
ONEOK	84	—	(4)	84	—	20
Pacific Enterprises	161	9	(26)	288	—	(171)
Peoples Energy Corp.	140	—	—	100	—	—
Piedmont Natural Gas Co.	50	—	—	100	—	—
UGI Corp.	93	—	—	57	—	43
Washington Gas Light Co.	98	—	—	100	—	—
Total and percent of total	$1,279	4%	(2%)	114%	—	(16%)

Source: Merrill Lynch & Co., Inc.

[a]Numbers reflect rounding.

ducer and resold it at the other end of the line to an LDC. Contractual clauses, backed by FERC orders, imposed a take-or-pay requirement at the producer–pipeline interface and also imposed a minimum bill between the pipeline and the LDC, which together guaranteed the flow. Take-or-pay clauses required purchasers to pay for a specified minimum quantity of gas regardless of whether they accepted delivery. The clauses served to protect producers with back-to-back deals. They also created a direct market discipline. The clauses effectively controlled deliverability in certain large gas-producing markets. During the 1960s and 1970s, reserves in the giant Hugoton field in Kansas, for example, had 30-year reserve lives but hardly any take-or-pay protection. Gulf Coast producers with strong take-or-pay clauses forced their gas on the market and effectively shut in older fields elsewhere.

Thus, under the old regime, surplus producing capacity remained submerged by the regulatory status quo. A "gas bubble" occurred in the 1970s and 1980s, but it was not obvious, because under the old gas contracts, the gas never left the ground.

Beginning in 1984, the old order began to fall apart when FERC eliminated minimum bills but allowed the continuation of take-or-pay clauses. This ruling imposed impossible market conditions on the industry and created a spot market almost overnight. Pipeline companies were legally committed to buying high-cost gas from producers, often at escalating prices, but they no longer had reliable markets for the gas, because their customers—LDCs and power plants—had switched to other, cheaper spot-rate gas suppliers. At about the same time, FERC changed

Table 2. Distribution of Assets by Activity—Major Companies by Industry Segment, 1991

Industry Segment and Company	Assets ($billions)[a]	Percent of Assets: Pipeline	Exploration and Production	Local Distribution	Refining and Marketing	Other
Producers/Gatherers/NGLs						
American Oil & Gas Corp.	$ 0.2	100%	—	—	—	—
ENSERCH Corp.	3.2	—	40%	43%	—	17%
Equitable Resources	1.4	—	51	54	—	(4)
MAPCO, Inc.	1.7	33	—	—	30%	37
Mitchell Energy & Development Corp.	2.3	16	41	—	—	43
Questar Corp.	1.2	31	28	36	—	5
Southwestern Energy Co.	0.4	—	54	43	—	3
Total and percent of total	$ 10.3	15%	33%	26%	5%	21%
Pipeline owners/operators						
Arkla	3.4	49	16	34	—	2
Coastal Corp.	9.5	49	8	—	32	10
Columbia Gas System	6.3	56	14	30	—	1
Consolidated Natural Gas Co.	5.0	30	28	40	—	1
El Paso Natural Gas	2.1	100	—	—	—	—
Enron Corp.	10.1	66	14	—	—	20
Panhandle Eastern Corp.	6.9	102	—	—	—	(2)
Sonat	3.2	47	31	—	—	22
Transco Energy Co.	4.6	72	9	—	—	20
Williams Companies	4.2	64	—	—	—	36
Total and percent of total	$ 55.5	63%	12%	9%	5%	11%
Large LDCs						
Atlanta Gas Light Co.	1.4	—	—	99	—	1
Brooklyn Union Gas Co.	1.7	—	5	95	—	—
MCN Corp. Holding Co.	1.5	—	—	92	—	8
National Fuel Gas Co.	2.3	—	9	80	—	11
NICOR	1.1	—	3	88	—	9
ONEOK	6.7	2	10	61	—	27
Pacific Enterprises	1.5	—	—	100	—	—
Peoples Energy Corp.	0.7	—	—	97	—	3
Piedmont Natural Gas Co.	1.0	0	0	42	—	58
UGI Corp.	1.0	—	—	42	—	58
Washington Gas Light Co.	1.0	—	—	100	—	—
Total and percent of total	$ 20.5	3%	6%	77%	—	14%

Source: Merrill Lynch & Co., Inc.

[a]Numbers reflect rounding.

rate designs, which made pipeline profits entirely dependent on moving throughput. Pipelines were supposed to provide utility services but often while competing with one another. Thus began an eight-year bout with a deregulatory agenda that made losers of producers, pipelines, and most of all, shareholders.

The pipeline companies' merchant function—buying natural gas at one end of the pipe and reselling it on a back-to-back basis at the other end—steadily disappeared as the new deregulated spot market grew. **Figure 1** shows the shift in the post-1984 gas market when gas pipelines tried to sell natural gas at $5 a thousand square feet (mcf) in a $2 market. The pipeline segment of the industry shifted from a market of 97 percent sales to a market of 90 percent transportation. When FERC Order 636 is fully implemented in November 1993, 100 percent of interstate pipeline volumes will be transportation alone. Pipelines have become contract carriers, and under Order 636, they will pass through all costs and surcharges in transportation rates.

The emerging organizational structure for integrated natural gas companies in the 1990s is shown in **Figure 2**. Gas flows from production to pipeline; this structure unbundles the gas business. Regulated pipelines and LDCs feed surplus cash flow into unbundled or deregulated subsidiaries, thereby controlling the flow of gas without owning the commodity. The structure of Enron is an example of this model, but because Enron does not have a distribution company, the gas goes directly into its power

Figure 1. Interstate Pipeline Throughput, 1982–92[a]

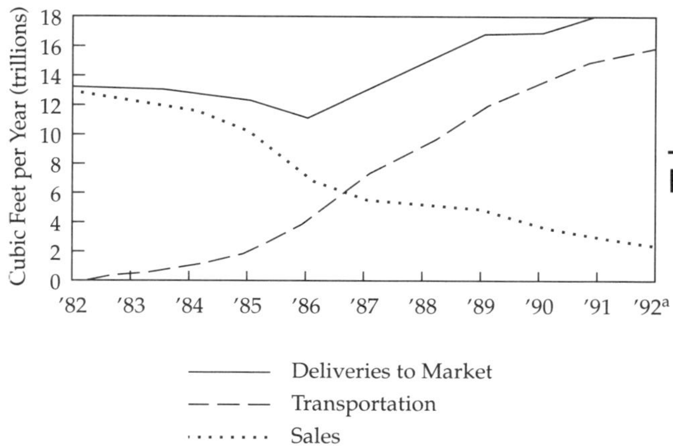

Source: Interstate Natural Gas Association of America annual surveys.

[a]Estimate.

plants on the Houston ship channel. Controlling the flow of natural gas was ideal in the 1980s; those companies that owned the reserves and produced the gas lost a great deal of money. Enron would today probably sell 99 percent of its pipeline assets on one condition—that it could continue to operate the pipeline systems. As long as it operates the systems, it controls the flow of natural gas.

Figure 3 shows a more diversified structure than the model in Figure 2. It is roughly similar to the structure of Coastal, where surplus cash flow moves into diversified energy businesses. This model was popular in the 1970s and early 1980s. Recent efforts to diversify have failed because of the tremendous deflation in the energy market from 1984 through 1992. History will probably repeat itself in the post-1995 era, however, when the industry's balance sheets become more liquid again and the urge to diversify in order to grow reasserts itself.

The Rate-Making Process

Regulated natural gas companies earn returns only on their rate base, not on the underlying commodity. Many investors think that if gas prices increase, so do pipeline profits, but under regulation, this is not true. Only in unregulated or lightly regulated intrastate contracts, which represent a tiny part of the overall gas market, are margins a function of the delivered price at the pipeline.

Everything important in regulated natural gas economics derives from the rate case, the process through which each company and regulatory agency negotiates rates. One function of the rate case is to allocate costs of service and returns between fixed costs and variable costs. Most of a pipeline company's costs are fixed costs related to operating the system; variable costs are incurred whenever the company moves volume, but they are usually a minute factor in rate making.

Rate Design

Rate making would be much simpler if the industry had a one-part rate system. The modified fixed/variable (MFV) rate structure that has been in general use is a two-part rate, however; thus, a rate case will assign fixed costs according to a two-part billing system. Two-part billing is similar to the billing

Figure 2. A Modern Integrated Natural Gas Company

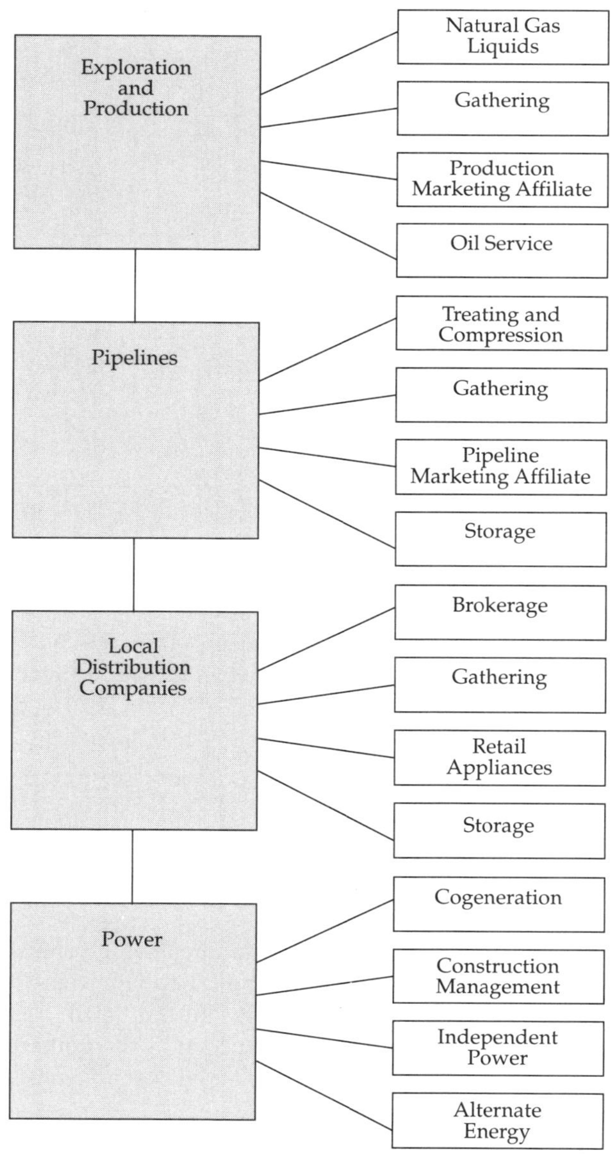

Source: Merrill Lynch & Co., Inc.

Figure 3. A Diversified Natural Gas Company

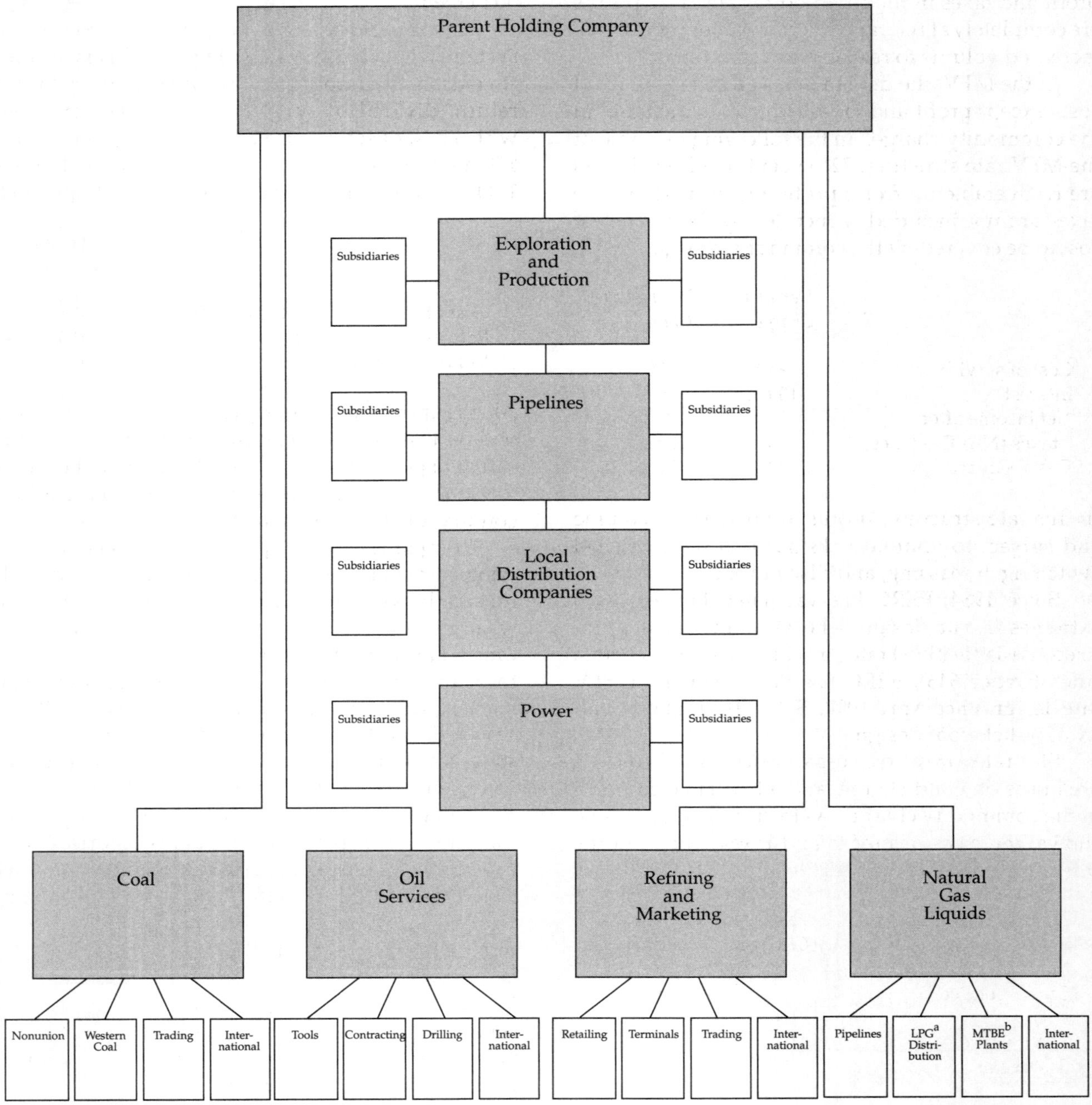

Source: Merrill Lynch & Co., Inc.

[a]Liquid propane gas.
[b]Methyl tertiary butyl ether.

system used by telephone companies, which charge a flat monthly fee whether the customer uses the service or not, plus a usage fee. In the natural gas industry, the front-end load, known as the demand or reservation charge, covers all fixed costs. Customers pay it irrespective of use.

Pipeline companies also charge a back-end load, the commodity charge, whenever volume moves. The gamesmanship in rate making occurs in the cost shifting related to the commodity charge. It is loaded or unloaded depending on where FERC dictates companies put profits and related income taxes. When pipelines are allowed to put profit and taxes in the front end, the demand charge, they realize 100 percent of allowable profits regardless of volume. This process is called "bulletproofing" and follows the classical utility model. If companies must put profit and taxes in the commodity rate, their profits are completely at risk, and the companies must move increased volume to realize allowed returns.

In the MFV, the demand charge covers all fixed costs except profit and taxes, which are included in the commodity charge. In the following example of the MFV rate structure, 72 percent of all fixed costs are recoverable up front; profit and related income taxes are not included, which leaves 28 percent of cost to be covered by the commodity charge:

	Demand Charge	*Commodity Charge*
Cost of service	95%	5%
Interest	100	0
Net income after taxes (NIAT) + taxes	0	100
Cost recovery	72%	28%

In this rate structure, profits are driven by volume and subject to volume risks such as weather, fuel switching, bypassing, and discounting.

Since 1984, FERC has vacillated between two extremes in rate design—the MFV and the straight fixed/variable (SFV) rate structure. Under the "final rule" (Order 636), FERC has been mandating SFV rate design since April 1992. SFV is the highest quality of bulletproof design.

SFV rates are also two-part rates. All fixed costs are in the demand charge, and all variable costs are in the commodity charge. As the following example shows, the costs of profits and taxes, unlike in the MFV, are front-end loaded:

	Demand Charge	*Commodity Charge*
Cost of service	95%	5%
Interest	100	0
NIAT + taxes	100	0
Cost recovery	97%	3%

In this system, pipeline companies recover 97 percent of all costs, including profits and taxes, in the monthly reservation charge paid by the customer or shipper. The demand charge plus the commodity charge equals the 100 percent load-factor rate, which is either the firm transportation rate (which applies to firm transportation contracts) or the interruptible transportation rate (which applies to interruptible contracts).

An alternative to the MFV and the SFV is an enhanced fixed/variable rate, which would be a compromise between the two extreme systems. The demand charge covers 50 percent of profits and taxes, and the commodity rate covers the remaining 50 percent.

An example of a one-part rate is the volumetric system, which is used in a variety of industry sectors. It consists of simply the cost of service, including return, divided by annual volume. For example, with a cost of service of $100 million and a return of $30 million, the volumetric rate based on volume of 1,000 billion cubic feet (bcf) would be $0.13 per mcf:

	Totals	*Per Thousand Cubic Feet*
Cost of service	$100,000,000	$0.10
Return	$ 30,000,000	0.03
Volumetric rate	1,000	0.13

All the intrastate pipeline companies in Texas use one-part rates. For the same $0.13 rate, companies can transport gas 5 miles down the road to the Houston ship channel, the largest natural gas market in the country, or 700 miles away to Midland, Texas.

Companies use one-part rates in interstate gathering to remain competitive. The rates are usually mileage based or zone based. They work well in cold weather when companies can exceed design volumes. Earnings can run into trouble, however, when there are volume shortfalls or new competition in the market. Discounting is frequent, and the results are seasonal. Williams Natural Gas has the only interstate transmission pipeline on a volumetric rate basis, and it is moving toward the SFV rate.

Filing cycles differ according to jurisdiction. At the federal level, pipeline companies historically had to redo the rate case at least once every three years. This requirement has effectively been nullified by Order 636. At the state level, many LDCs have annual redetermination of allowable returns on equity (ROEs). They may also have, depending on the state, supply-adjustment mechanisms for natural gas costs. Some states have three-year rate-case intervals; others do not require adjustment at all. Canada has annual redeterminations.

Costs and the Rate Base

All costs incurred on interstate contracts flow through on an as-billed basis by way of the federal preemption doctrine: If the costs are deemed prudent by a federal regulatory body and are allowed to be recovered, states and localities can do little to stop the process. The doctrine arose from the Supreme Court's decision in *Nantahala Power & Light Co. v. Thornburg, Attorney General of North Carolina, et al.*, and it is an important factor when dealing with take-or-pay, transition, or environmental costs—that is, these costs are passed from the pipeline company to the LDC to the customer. On some occasions, lawyers have challenged the doctrine, but they have been unsuccessful in the courts. Regulatory agencies in some states have also attempted to circumvent the process but with limited success. LDCs in Pennsylvania, for example, agreed to absorb about 10 percent of take-or-pay flow-through costs.

Take-or-pay surcharges are split between demand and commodity rates. Companies that discount those surcharges prolong their cost recovery. Transition costs under Order 636 will be important for certain pipeline systems. The costs will be recovered 100 percent out of the demand charge. The companies with the largest transition costs include Sonat and Texas Eastern, and thanks to Order 636, they should have no problem recovering those costs.

In addition to assigning costs, the rate case determines test periods, design volumes, and load factors. Test periods, design volumes, and cost data are often based on the latest 12 months of volume data. The load factor is the annual daily volume divided by peak-day volume. A pipeline system with 1 bcf a day of capacity running at a rate of about 600 million cubic feet a day has a 60 percent load factor. For rate making based on fixed costs, rates are designed to pay for 100 percent of the 1-bcf daily capacity, even though customers may use less. A 100 percent load factor usually sets the base rates in the rate case.

Exhibit 1 is a hypothetical rate-base model with itemized assets. The rate base consists of the gas utility's net plant, property, and equipment, plus gas storage, materials, and supplies, less deferred taxes, investment tax credits, and customer advances. (Deferred taxes apply to property accounts and accelerated depreciation. Deferred taxes and investment tax credits are common on LDC books; pipelines show deferred taxes from the property account.) The model yields a rate base of approximately $1 million.

The model uses a 50/50 capital structure with 45 percent debt, 5 percent preferred stock, and 50 percent common equity. Regulators at FERC, which has a rate-of-return staff, seem to favor this capitalization structure. At the LDC level, regulators tend to be slightly more generous; many jurisdictions allow as much as 60 percent equity.

Analysts can spend months trying to figure out what the rate-base returns should be based on complicated annual regulatory filings, or they can simply determine the capital structure, apply it against the rate base, and then figure out what kind of ROE the pipeline is earning. Both approaches yield the same result. The model in Exhibit 1 uses a 9 percent cost of debt, 10 percent cost of preferred stock, and 13 percent cost of common equity for a weighted-average cost of capital of 11.05 percent. The weighted-average cost of capital is used to calculate the return on the rate base—in this case $110,500.

Exhibit 1. Hypothetical Rate-Base Model

Rate base (thousands)	
Gas utility plant, property, and equipment (PP&E), net	$1,000
Gas storage	
Working gas	150
Base gas	100
Materials and supplies	50
Total	$1,300
Less:	
Deferred taxes, PP&E	(200)
Deferred interruptible transportation costs	(50)
Customer advances	(50)
	($ 300)
Resulting rate base	$1,000
Capital structure (thousands)	
Long-term debt	$ 450
Preferred stock	50
Common equity	500
Total	$1,000

Rate of return	Cost of Capital	Leverage	Weighted-Average Cost of Capital
Long-term debt	9.0%	45%	4.05%
Preferred stock	10.0	5	0.50
Common equity	13.0	50	6.50
Return on rate base			11.05

Rate-base returns (thousands)

Debt = 4.05% × 1,000 = $40.50 = 9% × $450

Preferred stock = 0.5% × 1,000 = $5 = 10% × $50

Equity = 6.5% × 1,000 = $65 = 13% × $500

Return on rate base = 11.05% × 1,000 = $110.50

The figure for gas storage is a function of sales, and regulated pipeline gas sales are rapidly disappearing. This loss creates a rate-base problem; in the example, as much as $250 million, 25 percent, may leave the effective rate base of $1 billion and move into some unregulated, nonjurisdictional area. As the jurisdictional rate base decreases, reported operating income and net income decline. Ideally, the more companies spend on increasing their rate bases, the more they will earn.

Because natural gas has become unbundled, suppliers and shippers now pay individual rates based on a menu of services instead of one flat rate. In the future, rates will be designed to reflect this unbundling. They will include 10- to 20-year firm transportation contracts that will permit buying peak-day capacity rights on the system with seasonal variation. Under Order 636, companies that do not need capacity can broker the excess by posting unused capacity on an electronic bulletin board, collecting the monies accordingly, and using that money to credit against their cost of service. The pipeline companies' alternative would be interruptible transportation contracts—spot capacity made available on a first-come basis either through its latest rate-case design volume or underutilized firm transportation volumes. Under Order 636, however, interruptible capacity may amount to only about 5 percent of rate-design volumes. Usually, interruptible transportation rates are designed to match firm transportation rates on a 100 percent load-factor basis. Storage service would also be available on a firm or interruptible basis.

Regulators usually examine the prior 12 months' costs and volumes when setting rates. All costs and volumes are based on original costs or historical book figures. These practices present two kinds of cost problems. First, companies cannot buy a pipeline for a 50–100 percent premium to rate-base equity and expect to recover that premium out of rates. An estimated $7–$9 billion of goodwill appears on pipeline books today as a result of the acquisition spree of the mid-1980s. Pipeline acquisitions in the past also were burdened by retrospective rate making and ROE attrition. Second, rapidly growing pipelines or LDCs can be squeezed, because depreciation, interest expense, and operating costs increase faster than the companies are able to increase rates. For example, Washington Gas Light has enjoyed some of the most rapid growth in the LDC industry; its rate base is growing 9 percent annually, but its earnings per share have been growing at only 2–3 percent annually.

The Canadian system, which is tough but very fair, uses prospective rate making. Companies forecast cost of service, interest expense, spending, and equity needs for the next year and include them in the rate case. As a result, TransCanada and Westcoast Energy never had the take-or-pay or restructuring problems the U.S. operators had, and they now have beautiful pipeline systems.

Some states allow prospective rate making. The Georgia Public Service Commission, for example, reluctantly adopted prospective rate making at the insistence of the legislature, and even New York, one of the hardest places to make money in a regulated industry, partially uses prospective rate making. Some states offer incentive rate making. (Michigan, for example, has been proactive in this area.) Incentives have allowed companies with 12–13 percent ROEs to earn ROEs of 15–16 percent by sharing efficiency gains.

Another aspect of rate making is the concept of value of service, which refers to the LDC practice of buying natural gas from a pipeline company and reselling it at prices equivalent to fuel oil prices. If gas prices are low and fuel oil prices are high, LDCs can share the difference with their customers. UGI in Pennsylvania is the only company in the industry smart enough to have a value-of-service provision built into its rate case.

The rate base and returns are asset driven. As Exhibit 1 shows, capital structure is used to create rate-base equity. Regulators typically apply a 50/50 debt-to-equity ratio against the rate base to determine the rate-base equity, and if the public book capital structure differs too much from regulatory perceptions, the regulators will impute their own. Rate-base equity often exceeds book equity because the industry has taken a tremendous number of book write-offs as a result of take-or-pay and other restructuring charges. Since 1985, the industry has collectively written off $3.5 billion pretax, $2.5 billion after tax, of net worth to absorb take-or-pay charges. Because rate-base equity can be much higher than reported book equity, transferring a 13 percent return on rate-base equity to the net income shown in the public books can often result in an inflated ROE.

An extreme example of this situation is Columbia Gas System, which is now in Chapter 11 bankruptcy because 3 percent of its suppliers refused to renegotiate contracts in 1985. In other words, these producers took the company under. The company currently has a negative net worth of about $700 million but a rate base of about $1.6 million. A capital structure of 60 percent debt and 40 percent equity gives the company a rate-base equity of $600 million. Multiplying the rate-base equity by an approved ROE of 13 percent gives Columbia Gas earnings of $78 million a year at SFV rates. The company now shows $785 million of cash on the books because its transmission company generates so much cash.

Regulators usually use discounted cash flow models to figure allowed returns. This is the point at which interest rates play a role in rate making. DCF models can be manipulated to prove just about anything, which is exactly what regulatory rate-of-return staffers try to do. The result is extremely low allowed returns; the current range, based on the 50/50 capital structure, is between 11.6 and 13 percent. Note, however, that "earned" ROEs should improve after 1993 because of the bulletproofing effect of Order 636.

In **Figure 4**'s comparison of yields on 30-year T-bonds and the ROEs of public utility commissions, note that the spreads widened in the mid-1980s and

Figure 4. Public Utility Commission ROEs versus Yields on 30-Year Treasury Bonds, 1980–3Q'92

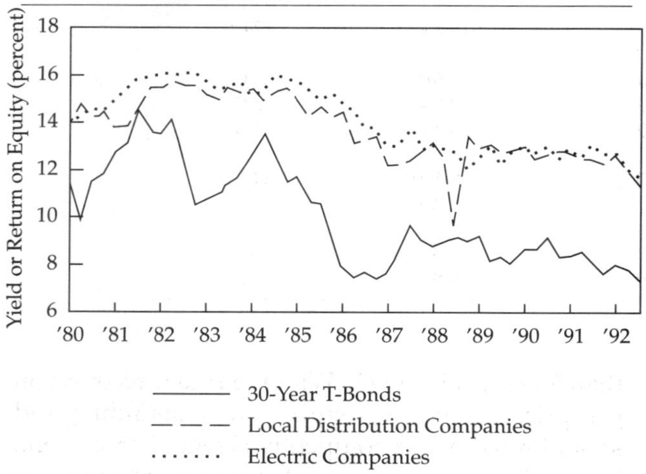

Source: Merrill Lynch & Co., Inc.

Note: ROEs are the average ROE weighted by amount granted.

appear to have stayed there. Allowed returns have been under pressure because of declining interest rates, and LDC returns are showing a gradual decline.

Effects of Order 636

Columbia Gas and Transco Energy had the first SFV rate cases approved in April 1992, but most companies will have moved away from MFV and phased in SFV rates by November 1, 1993. As they move to SFV rate structures, the companies' operating profit patterns will shift. **Figure 5** compares profits under the two rate structures.

Figure 5. Annual Operating Profit Patterns—MFV and SFV Rate Structures

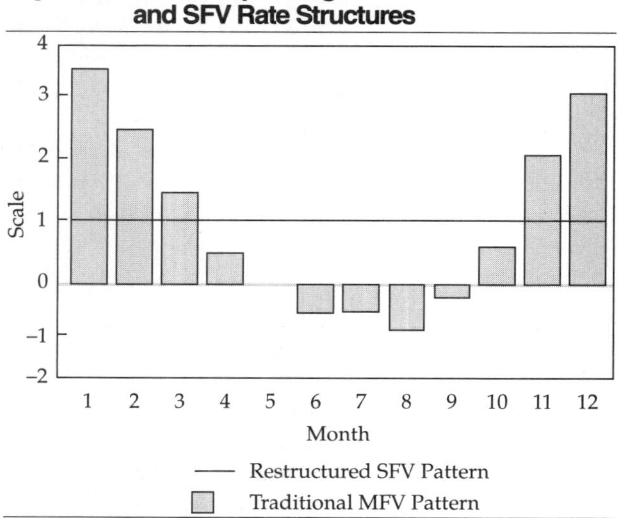

Source: Williams Companies.

Under SFV rates, profits are constant, which provides relief for the pipeline companies that have for years been underearning under the traditional MFV design. Of course, companies that have overearned in the past may not be able to do so in the future; they will have to figure out other ways to overearn.

Under the SFV structure, profits will not be at risk from bypassing or fuel switching. Moreover, because they will recover all fixed costs and profits up front, companies will not get injured by rate discounting. Discounting shifts to the aftermarket; the shipper that has unused capacity and wants to resell it will accept a discount.

SFV will also reduce the weather sensitivity of earnings. **Figure 6** illustrates the pattern of Williams Companies' operating profits created by a combination of the volumetric system for setting rates (which has the same weather sensitivity as the MFV system) and unusually warm winters in 1987 and 1990. If Williams had been regulated by the SFV system, it would have experienced the slow and steady growth of the columns.

Table 3 provides an example of the effect of Order 636 on rates. If Texas Eastern moves from the old rate to the proposed SFV rate, the front-end load will increase 74 percent in the monthly reservation charge. Transition cost surcharges will increase the front-end load by 21 cents. The commodity rate will drop 84 percent, from 32 cents to 5 cents. Overall, customers buying natural gas from Texas Eastern, instead of paying 35 cents up front in the demand charge, will pay 87 cents up front.

Table 4 shows the impact of Order 636 on cost of

Figure 6. Weather Sensitivity of Operating Profits—Volumetric and SFV Rate Systems, 1987–91

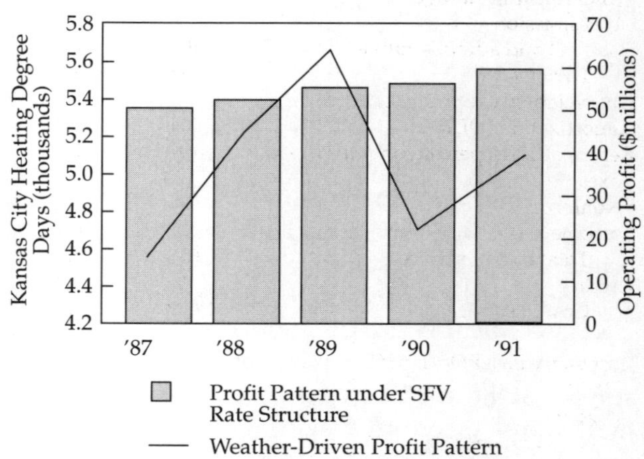

Source: Williams Companies.

Note: In a normal year, the number of heating-degree days (see Glossary, pages 151–53) is 5,283.

Table 3. Order 636 Rate Impacts—Texas Eastern

Type of Contract	Old Rates	Proposed SFV Rates	Percent Change
Firm transportation (FT)			
Monthly reservation charges	$10.775	$18.750	74%
Conversion to daily basis			
(30.4 average days/month)	0.350	0.610	74
Commodity rate	0.320	0.050	(84)
100 percent load-factor rates	0.670	0.660	(1)
Interruptible transportation (IT)	0.670	0.600	(10)
Transition cost surcharge	—	0.210	
Total FT	0.670	0.870	30
Total IT	0.670	0.810	21

Source: Texas Eastern Transmission Co. rate-case filings.

service. Even though gas costs decline drastically as companies move out of the merchant function, the return is virtually the same—$188 million versus $190 million.

Industry Returns

Return on equity is the accepted performance measurement for investors in the natural gas industry. As shown in **Figure 7**, returns for producers started strong in 1980 at 23 percent but then declined to less than 5 percent by 1990. With the recent recovery in gas prices, however, returns are rebounding and should recover to a competitive level of 15 percent. Fuel switching in key markets may provide a ceiling in the overall price trends by 1993 or 1994.

Returns for pipeline companies did not fare much better than those for producers. Pipeline ROEs collapsed in about 1983 when the change in rate designs eliminated minimum bills and created the spot market. These companies' returns are also rebounding, however, and should improve in 1993 and 1994, mainly because of improvements in rate designs.

Not all companies lost during the past decade. For most of the 1980s, the LDCs were buyers in a buyers' market. Returns were relatively high and stable except in 1988 and 1989, when they dipped because of warm weather. (Falling interest rates also

Table 4. Order 636 Cost-of-Service Impacts—Texas Eastern
(millions of dollars, except as noted)

	Old Rate Plan	Proposed SFV Rate Plan
Operating and maintenance (O&M)		
Production, gathering,		
and extraction	$ 12	$ 3
Gas costs	848	0
Other gas supply expense	2	0
Underground storage	43	39
Transmission	274	235
General and administrative	108	105
Total O&M	1,288	382
Depreciation, depletion, and		
amortization (DD&A)	101	101
Return at 12.13 percent on rate base	190	188
Taxes		
Nonincome	36	36
Income	67	66
Total cost of service	1,681	772
Credits	(19)	(18)
Cost of service	1,662	754
Cost profile under Order 636		
O&M		49%
DD&A		13
Taxes, nonincome		5
Direct cost of service		67
Return + income taxes		33
Total cost of service		100%

Source: Texas Eastern Transmission Co. rate-case filings.

Figure 7. Return-on-Equity Profiles, 1980–90

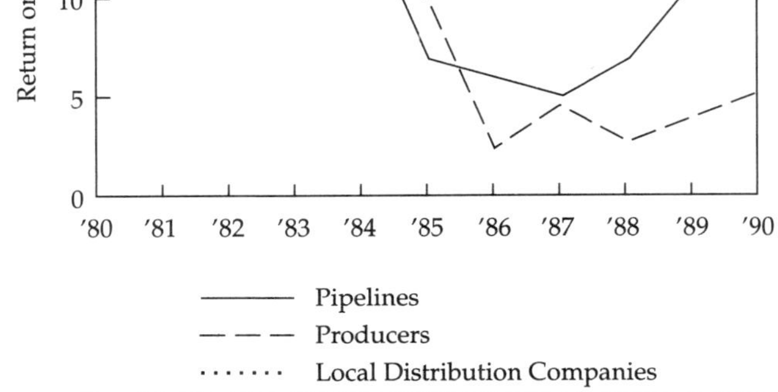

Sources: Shell Oil Co.; American Gas Association; Federal Energy Regulatory Commission Form 2 filings for 1980–90.

put pressure on allowed ROEs.) The way in which state regulators dealt with changing markets and applied the test of "just and reasonable return" to the LDCs they regulate was more favorable to LDC returns than the way in which federal regulators dealt with pipeline companies. Thus, LDC returns stayed at decent levels.

Table 5, which shows summary 1990 results from Form 2, an annual filing with FERC, indicates that some pipeline companies also could earn generous ROEs in the decade because they moved tremendous volume—more than figured into the rate-case calculation. (Note that the ROE figures in Table 5 represent only those returns for regulated pipeline subsidiaries, not for consolidated companies.) Because profitability is a direct function of volume under the MFV rate design, companies such as Northern Natural Gas, ANR Pipeline, and Northwest Pipeline earned higher ROEs by moving large incremental volumes than would have otherwise been possible.

Diversification hurt the pipeline industry. **Table 6** shows the ROEs of the pipeline companies as reported in their public books. In the 1984–92 era, these companies earned a mere 6.6 percent return on their public book equity. Generally speaking, returns from the regulated area (normalized at 13 percent) were measurably better than diversification returns, some of which were bona fide money losers.

Valuation

The quality of rate design is improving, and so is the quality of earnings for pipeline companies and LDCs. Interest rate declines, however, often translate into declines in allowed ROE, which stunts rate-base growth and flattens earnings.

Industry analysts should look at a natural gas

Table 5. Return on Equity—Major Pipeline Companies, 1990
(millions of dollars, except ROEs)

Company	Gas Utility Operating Income (GUOI)	Interest Expense	GUOI less Interest Expense	Common Equity	Return on Equity
Algonquin	$ 24.2	($ 9.7)	$ 14.5	$ 114.3	12.7%
ANR Pipeline	175.4	(58.7)	116.6	927.9	12.6
Arkla	146.7	(163.3)	(16.6)	1,224.7	NM
CNG Transmission	89.5	(33.9)	55.5	472.2	11.8
Colorado Interstate	77.4	(24.8)	52.6	422.5	12.5
Columbia Gas System	81.5	(104.8)	(23.3)	164.9	NM
Columbia Gulf	18.3	(4.4)	13.8	172.9	8.0
East Tennessee	10.4	(2.4)	8.0	60.1	13.4
El Paso Natural Gas	140.7	(99.4)	41.3	1,769.7	2.3
Florida Gas Transmission	26.2	(3.6)	22.6	126.6	17.9
Great Lakes	43.9	(17.8)	26.1	208.7	12.5
KN Energy	22.0	(16.9)	5.1	181.8	2.8
Mississippi River Transmission	15.6	(6.1)	9.5	95.9	9.9
Natural Gas Pipeline	177.9	(21.5)	156.3	1,144.1	13.7
Northern Border	96.5	(40.2)	56.3	246.7	22.8
Northern Natural Gas	96.0	(6.6)	89.3	519.7	17.2
Northwest Pipeline	78.6	(20.8)	57.8	304.5	19.0
Panhandle Eastern Pipeline	65.0	(129.1)	(64.1)	690.7	NM
Southern Natural Gas	83.2	(36.7)	46.6	631.7	7.4
Tennessee Gas Pipeline	164.2	(280.9)	(116.7)	6,314.6	NM
Texas Eastern Transmission	135.4	(90.9)	44.5	1,126.0	4.0
Texas Gas Transmission	58.0	(31.4)	26.5	352.0	7.5
Transcontinental Gas Pipeline	139.2	(78.7)	60.5	480.3	12.6
Transwestern	45.0	(13.1)	31.9	289.9	11.0
Trunkline	22.3	(32.1)	(9.7)	942.4	NM
United Gas Pipeline	101.3	(54.0)	47.3	208.5	22.7
Williams Natural Gas	13.8	(18.5)	(4.6)	216.5	NM
Total	$2,148.1	($1,400.4)	$747.8	$13,745.3	5.4%

Source: Federal Energy Regulatory Commission annual Form 2 filings.

Note: NM = not a meaningful figure.

Table 6. Corporate Return on Equity of Major Diversified Pipeline Companies, 1984–91
(millions of dollars, except ROEs)

Company	1991	1990	1989	1988	1987	1986	1985	1984
Net income								
Arkla	$5.6	$91.1	($68.7)	$117.0	$110.0	$80.0	$100.0	$69.0
Coastal Corp.	95.8	225.1	177.8	153.7	97.3	38.0	113.7	97.4
Columbia Gas System	(694.4)	104.7	145.8	111.1	100.5	87.9	(107.0)	165.5
Consolidated Natural Gas Co.	168.6	163.8	181.7	192.9	186.0	178.7	216.2	211.6
El Paso Natural Gas	143.0	187.0	165.0	NAp	NAp	NAp	NAp	NAp
Enron Corp.	217.0	177.2	226.0	130.0	54.0	373.0	116.0	230.0
Panhandle Eastern Corp.	85.1	(233.4)	69.5	59.0	109.7	114.5	101.8	131.3
Sonat	77.9	91.2	95.6	65.6	70.1	(115.8)	65.8	161.9
Transco Energy Co.	(193.1)	15.4	88.7	(91.0)	(75.5)	58.3	19.1	128.2
Williams Companies	98.4	65.4	44.8	14.2	53.7	(120.4)	31.6	34.9
Total	($686.2)	$1,430.5	$1,692.2	$726.5	$644.9	$880.2	$632.1	$1,370.8
Common equity								
Arkla	$886.2	$985.4	$579.9	$710.0	$688.0	$791.0	$768.0	$750.0
Coastal Corp.	2,029.7	1,968.9	1,780.9	1,259.8	730.1	659.2	642.3	447.9
Columbia Gas System	1,006.9	1,757.8	1,620.3	1,552.6	1,523.7	1,448.7	1,422.7	1,634.2
Consolidated Natural Gas Co.	1,889.8	1,844.6	1,671.9	1,634.8	1,598.2	1,545.7	1,482.4	1,359.9
El Paso Natural Gas	1,815.0	1,828.0	1,715.0	NAp	NAp	NAp	NAp	NAp
Enron Corp.	1,706.5	1,618.0	1,548.0	1,399.0	1,373.0	1,442.0	1,225.0	1,442.0
Panhandle Eastern Corp.	1,333.1	1,137.8	1,415.1	797.7	1,029.8	991.8	1,251.7	1,217.4
Sonat	1,042.7	1,060.5	1,035.3	1,010.2	1,023.0	1,005.9	1,276.3	1,301.5
Transco Energy Co.	386.8	600.5	606.4	534.2	650.1	718.3	726.5	872.2
Williams Companies	1,070.0	1,016.5	998.3	897.8	831.8	614.3	1,212.8	1,228.5
Total	$15,934.6	$17,185.0	$16,248.1	$12,957.0	$13,220.7	$13,628.9	$14,922.7	$15,591.6
Returns on equity								
Arkla	0.6%	9.2%	(11.9)%	16.5%	16.0%	10.1%	13.0%	9.2%
Coastal Corp.	4.7	11.4	10.0	12.2	13.3	5.8	17.7	21.7
Columbia Gas System	(69.0)	6.0	9.0	7.2	6.6	6.1	(7.5)	10.1
Consolidated Natural Gas Co.	8.9	8.9	10.9	11.8	11.6	11.6	14.6	15.6
El Paso Natural Gas	7.9	10.2	9.6	NAp	NAp	NAp	NAp	NAp
Enron Corp.	12.7	11.0	14.6	9.3	3.9	25.9	9.5	16.0
Panhandle Eastern Corp.	6.4	(20.5)	4.9	7.4	10.7	11.5	8.1	10.8
Sonat	7.5	8.6	9.2	6.5	6.9	(11.5)	5.2	12.4
Transco Energy Co.	(49.9)	2.6	14.6	(17.0)	(11.6)	8.1	2.6	14.7
Williams Companies	9.2	6.4	4.5	1.6	6.5	(19.6)	2.6	2.8
Industry average	4.3%	8.3%	10.4%	5.6%	4.9%	6.5%	4.2%	8.8%

Source: Annual reports.

Note: NAp = not applicable.

company's ability to pass through all fixed costs and to earn or exceed the allowed ROE. They should focus on assets, even though management often presents them with meaningless operating income information. Much of what analysts need to know is not contained in the company's annual report or even in the 10-K form; it is found in the rate case, which can be obtained from the company. In this connection, when visiting the company, the analyst should ask to speak to the people who work with the rate case and look at their work sheets. Usually, the rate-of-return calculations, rate base, and cost of service will all be in one volume; the rest of the volumes will be noise.

Consolidated financial statements are often meaningless to the valuation process, because asset profiles do not match regulatory or real-world profiles. Capital structures can differ drastically among subsidiaries. Goodwill, in accord with SEC accounting rules, is frequently pushed down, and analysts never see it. Both Enron and Occidental Petroleum have $2 billion worth of goodwill on their natural gas pipelines, but analysts never see it because of SEC rules. Diversification compounds the problem.

As an example of the relative importance of the financial statements and the rate case to investment

Table 7. Financial Data, Sonat and Southern Natural Gas, 1991
(millions of dollars, except as noted)

	Sonat		Southern Natural Gas		
	Annual Report	10-K	10-K	Form 2	Rate Case
Net plant	NAv	$817.00	$777.00	$592.00	$580.00
Gas storage	$122.00	122.00	122.00	186.00	190.00
Inventories	27.00	27.00	27.00	24.00	27.00
Other net working capital	NAv	NAv	NAv	NAv	$4.00
Total assets	NAv	966.00	926.00	802.00	801.00
Deferred taxes (property)	NAv	NAv	(106.00)	(109.00)	(115.00)
Estimated rate base	1,334.00	NAv	820.00	693.00	686.00
Capital structure					
Long-term debt	1,315.00	1,315.00	340.00	300.00	400.00
Common equity	1,043.00	1,043.00	656.00	656.00	513.00
Total	2,358.00	2,358.00	996.00	956.00	913.00
Debt	55.80%	55.80%	34.10%	31.40%	43.80%
Equity	44.20%	44.20%	65.90%	68.60%	56.20%
Total	100.00%	100.00%	100.00%	100.00%	100.00%
Cost of capital					
Debt	NAv	NAv	10.70%	9.96%	9.24%
Equity	NAv	NAv	11.41%	11.41%	15.75%
Total	NAv	NAv	11.17%	10.96%	12.90%
Rate-base equity	NAv	NAv	$540.00	$475.00	$386.00
NIAT[a] at 13 percent ROE			70.00	62.00	50.00
Tax effect at 37 percent			41.00	36.00	29.00
Net income before taxes			111.00	98.00	79.00
Rate-base interest			37.00	30.00	37.00
Net operating income (NOI) = pretax return			$148.00	$128.00	$116.00
Reported NOI	$140.00	$140.00	$136.00	$140.00	$110.00

Source: Sonat, Inc., public data.

Note: NAv = data not available.

[a]Net income after taxes.

analysis, **Table 7** compares Sonat's annual report and 10-K with the Form 2 and rate-base calculations for its pipeline, Southern Natural Gas, for 1991. Note that the annual report tells an analyst nothing about the rate base; nor does it show a net plant figure or deferred taxes. Gas storage and inventories are insignificant numbers. Although the ratios provided may be helpful, the annual report does not provide direct information about the regulated business's capital structure. In addition, the annual report's asset numbers are inflated with miscellaneous accruals or goodwill.

The 10-K provides better numbers than the annual report, but total assets appear as $966 million compared with $801 million listed in the rate case. Form 2 annual reports to FERC generally provide fairly good asset numbers—in this case, $802 million versus $801 million in the rate case. The Form 2 falls short on capital structure calculations, however. In the rate case, the capital structure is $400 million of debt and $513 million of equity (a 44/56 structure); Form 2 shows a completely different set of numbers and a 31/69 percent debt-to-equity structure. That discrepancy makes a big difference when net operating income is calculated.

In both the annual report and 10-K, the company reported $140 million in net operating income; in the rate case, it reported $110 million—a discrepancy of $30 million. The reason is that Sonat's reported numbers include those of Sea Robin, another of its pipelines. This practice is common among natural gas firms, so analysts must look carefully to see the entire picture.

Analysts can use various tools to develop company valuations. The most effective valuation tool for natural gas companies is a combination of price-to-earnings ratios and yield used together with ROE and balance sheet leverage. Although management can manipulate the books somewhat, analysts should filter out nonrecurring earnings or charges.

Cash flow from operations is a much-abused valuation device, and in the gas industry, it tends to be highly unpredictable. Cash flow from operations is defined as net income plus depreciation, depletion, amortization, deferred taxes, and other noncash items. Adjusted cash flow includes exploration expenses and recoveries of take-or-pay and transition costs. In this old-fashioned definition of cash flow, the greater the leverage, the more the cash flow. Typically, more assets add more depreciation, depletion, and amortization (and, presumably, more earnings) to cash flow accruals. In the pipeline industry, adding assets often does not work this way; diversification mistakes have created losses and deferred tax credits. Gas purchase costs for LDCs and transition costs for pipelines and LDCs can create massive swings in deferred taxes that make cash flow forecasting difficult. In addition, diversified natural gas companies have cash flow streams based on different depreciation and amortization schedules; for example, exploration and production assets can have an 8-year reserve life, pipeline assets a 50-year life.

Book value is a worthwhile valuation tool, because regulators use it when they set rates based on historical costs. Analysts should use it in tandem with other tools, such as the break-even transaction price (BETO), which is a measure of the maximum price a cash buyer can pay without diluting earnings. The BETO is calculated in such a way that incremental earnings will equal incremental after-tax interest and goodwill amortization.

Question and Answer Session

John E. Olson, CFA

Question: Are pipeline companies likely to discount demand charges?

Olson: Discounting was common under the MFV rates. Future discounting will depend on the relative bargaining posture of the pipeline in its particular market. SFV rates carry little incentive to discount either the demand or the commodity charge. All discounting should happen at the customer level in the aftermarket. For the record, some discounting of demand charges has occurred in recent years, but it remains a gray area in the minds of regulators and managers.

Question: Why is the cost of debt in the weighted-average cost of capital not an after-tax number?

Olson: The cost of debt is pretax because of regulatory convenience. It is used for rate-making purposes in rate-case calculations. Analysts can decompose capital structure numbers and turn everything into a pretax basis. In fact, most rate-of-return calculations today swing everything into a pretax return on rate-base calculations.

Question: How are the differences between the real cash earnings set by the rate case and the profits reported in the 10-K reconciled in the cash flow statements?

Olson: The difference shows up in a variety of ways—first, in depreciation and depletion adjustments; second, in deferred tax accruals; and third, in various reserves.

Question: Please describe the take-or-pay disasters in the context of rate regulation.

Olson: To deal with the take-or-pay problem, pipeline companies negotiated contract settlements with their customers under varying mechanisms that were ultimately combined into Order 528 in 1987. It was a disaster because the pipelines paid $10 billion to producers to buy themselves out of these contracts. They had to absorb about 41 percent, nearly $4 billion, of that cost. Most companies were able to recover the remaining $6 billion fully; they charged it to customers in the form of commodity rate surcharges or demand surcharges and recovered it over a negotiated period, typically three years. Columbia Gas had the biggest and most unique regulatory disaster. It had a binding legal agreement effective April 1, 1985, that specifically precluded any future litigation or possible extra cost recoveries on 97 percent of its supply contracts. The 3 percent of those supply contracts that were still outstanding, however, eventually forced the company into bankruptcy.

Question: What are Order 636's transition costs, and how do you quantify them? Also, what competitive advantages do the pipeline companies have as marketers?

Olson: Transition costs will include costs for restructuring remaining supply contracts (10 percent of volume in 1992); conversion costs incurred in moving to a transportation-only mode (i.e., gathering, NGL plant costs, electronic bulletin boards); and expenses for new receiving and delivery point hookups.

Pipelines should continue to enjoy market power in their new marketing role because of their years of hands-on experience. This expertise is worth plenty.

Question: Does the theory that FERC sets the upper bound and the competitive market sets the lower bound on earnings disappear under Order 636? Why won't pipeline companies have to discount the demand charge to maintain loads?

Olson: Order 636 will make the unfortunate dichotomy between regulation and market disappear. Pipelines won't need to discount demand charges under 636 because they become insensitive to volume and won't need to. Discounting shifts to the customers' unused capacity in the aftermarket.

Question: Please comment on the industry's performance outside North America, particularly in East Asia and Europe. How have the natural gas companies fared in the absence of regulations similar to those in the United States?

Olson: Overseas pipeline systems are largely government owned and have generally been excellent generators of foreign exchange, not to mention quite profitable investments. Continental Europe's regulation of pipelines has been protective and benign. Third-party access (or open-access transportation) has already arrived in the United States with Order 636, but it is perhaps 5–10 years away in Europe.

Question: Is Order 636 better for pipeline company bond performance than for pipeline equity performance? What is the next big problem or major issue facing pipelines?

Olson: Order 636 should affect both pipeline bonds and pipeline stocks favorably. The coming challenges will include getting final judicial endorsement of Order 636, avoiding any rate-case filings while the current costs of debt and equity capital prevail, and continuing to avoid any political burdens such as taxes based on Btu equivalency.

Perspectives on Natural Gas Pricing

Carol Freedenthal
Chief Executive Officer
Jofree Corporation

To forecast with any accuracy in the natural gas industry requires analysts to understand the basics of the industry and its pricing philosophy and mechanisms. This understanding involves a close look at such factors as fuel sources, the structure of markets, and trends in consumption.

The natural gas industry can be a topsy-turvy business. In 1992, prices were less than $1 during January and February and peaked in the summer at about $2.70. This inverted pricing trend makes forecasting difficult. But understanding the basics of the industry, as well as the philosophy and mechanics of pricing in the gas business, allows analysts and suppliers to look at the industry with some degree of accuracy.

Major Fuel Sources

The energy industry as a whole accounts for about 8 percent of the United States' $5.7 trillion GNP. Natural gas accounts for a bit more than 1 percent of GNP, or approximately $70 billion at the consumer level. Petroleum products are the leading fuel source in the United States, with natural gas and coal almost tied for second. Many people believe coal is dead, but as shown in **Figure 1**, it is not. Some people suggest that environmental laws will weaken coal's position, but investment in the proper equipment can make coal environmentally acceptable. In addition, the coal industry lobby is very strong, so the use of coal will not diminish to any great extent.

Petroleum use has declined since peak consumption in the late 1970s and will continue to decline. The United States has limited petroleum resources, and continued imports of oil and oil products will be expensive. The impact of the decline is not as bad as it appears, however, because foreign oil sources have moved into consumer markets in the United States and have much at stake in maintaining a good sales relationship with this country. Also, on a dollar basis, the decline is considerably less than on the basis of oil volume.

Natural gas, used mainly in home and space heating, accounts for about 25 percent of the fuel used in the United States. Natural gas receives more attention from Congress than any other fuel, because it affects average Americans who live in houses and work in factories. Congressional leaders are concerned about whether their constituents are satisfied with the price they pay for natural gas.

Industry Basics

Structurally, the natural gas business is relatively simple. Gas comes from two types of wells—*associated* wells, which produce oil and gas, and *gas-only* wells. In the early days of the business, almost all gas came from associated wells. Gas-only wells were either flared or capped.

With both gas-only and associated wells, the gas can go through a processing plant or directly into a pipeline. From a pipeline, it can go to a distribution system or directly to a consumer. All natural gas is transported from wellhead to consumer through pipelines. The transportation system allows a multitude of points at which to price natural gas. Thus, a quoted price could be the price at the wellhead, for example, or at the outlet pipe from the field gathering system, at the inlet into an interstate or intrastate pipe, at the city gate, or at the burnertip. When someone says the price today is $2.40 per million British thermal units (mmBtu), ask where that price is being charged: At the wellhead? At the inlet to the pipe? In Louisiana? Texas? New Mexico?

Discussions about price require care, and analysts must keep perspective when referring to gas prices. They must consider whether the price is a contract or cash price. If it is a contract price, is it short or long term, and is it a cash or futures price?

Figure 1. Total U.S. Energy Consumption, 1973–91

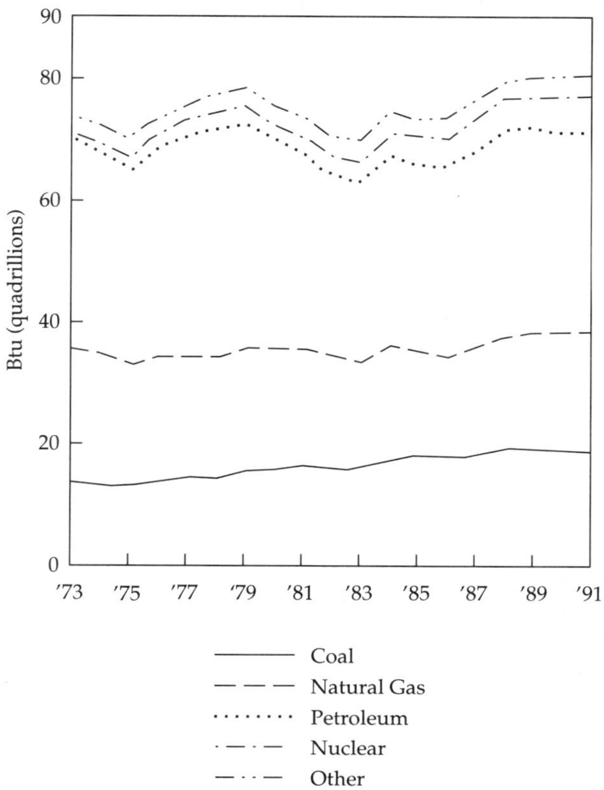

Source: Jofree Corporation.

Gas is seasonal, so companies sometimes use the annual average price in discussions. If they say, however, that prices are $2.40 per mmBtu for new gas at the wellhead, they most likely do not mean that it is $2.40 for the whole year. They mean that, for today, or for this month, it is $2.40 per mmBtu.

Natural gas as a product can be divided into two general markets. The first market is gas sold as a premium or preferred fuel. In the short run, premium-fuel customers are not going to switch energy sources, for several reasons. Once people have installed gas heating in their homes, for example, they will not tear it out because gas becomes scarce or the price climbs. They will retain gas heating. Certain industrial applications must also use natural gas. In the fertilizer industry, for example, ammonia plants must use natural gas, because it is the only feedstock in the United States for making ammonia. Fuel oil could be used as fuel to operate the plants, but the plants must use gas as a feedstock. Similarly, paper plants need natural gas as fuel for finishing the papermaking process; otherwise, the paper will be covered with black specks.

The secondary, or commodity, market for natural gas developed because of capacity in excess of demand for premium fuel. In this market, consumers can switch to alternative energy sources rather easily, so gas prices must be competitive. Gas competes with coal and residual fuel oil for certain industrial applications, such as boiler fuel and electricity generation. Its use in these applications is purely a function of price. Utilities will switch fuels to capture a 5- to 10-cent differential on 1 mmBtu. Florida Power and Light will change its fuel based on a 1-cent differential. It has fuel oil waiting in storage and natural gas available in the pipeline. If gas goes up a penny more than the fuel oil costs, the plant switches to fuel oil, and vice versa.

This dual-pricing situation encompassing premium and commodity markets precludes natural gas from reaching the maximum price. Natural gas is a premium fuel only if it is used in a premium application. Natural gas burned in a boiler, for example, is not premium fuel. Unfortunately, the marketplace does not announce whether it is using the fuel as a premium or a commodity product. If the price of gas goes above $2.60 per mmBtu in today's low-priced crude oil market, gas begins to lose commodity market share to less expensive residual fuel oil. The benefit of the higher price is lost as lower volumes of gas are consumed. On the other hand, if the price falls too low, producers have no incentive to drill new wells and develop more supply to meet increased demand, which creates, in effect, a lower limit on price. Thus, the "gas bubble," or surplus of available supply, does not dissipate totally. The gas bubble may diminish in impact, but it is always there. If gas producers lose market share for their fuel, the gas surplus grows and the price declines. If they cannot get the supply to meet demand when the price is low, they lose market share to competitors who can meet demand.

In summary, pricing—whether determined by the competition, commodity prices, or transportation costs—drives the natural gas market. If the price is not sufficiently high, producers will not look for new supplies or drill new wells. If the price is too high, gas will lose market share to competing fuels.

Another important force in the gas industry is uncertainty. The business is subject to uncertainty because of government regulation as well as supply and availability. Gas may be subject to price regulation at the wellhead, price regulation in transportation, and price regulation at the burnertip. Uncertainty resulting from actual or potential regulation directly affects the value of the gas industry.

Structure of the Industry

The U.S. natural gas industry as we know it today started in the early 1930s, when engineers developed seamless steel pipe that could withstand high pressures and move gas from southwestern to northern and northeastern gas markets. **Figure 2** shows how natural gas moved from the field to the consumer in the past. The old gas business revolved around the pipeline. The producer took the technical risk of drilling a hole in the ground, which until recently was a big risk because nobody knew whether anything was there two to five miles down. Pipeline owners became the merchants and financial risk takers, because southwestern producers had no desire to go north and negotiate the price of gas. The local distribution company (LDC) or the consumer played a role, but the pipeline assumed full financial responsibility.

No matter who bears the risk, natural gas in the United States can move only inside a pipe. From a marketing standpoint, this limitation is a major disadvantage for gas when compared with crude oil and coal. Crude oil, for example, can be moved in a tank truck or in a barrel as well as in a pipe. With natural gas, if the pipe does not reach the location of the gas, the producer cannot move the gas to market.

The ability to move gas directly from producer to consumer has been accomplished only in the past 10 years. The Natural Gas Policy Act (NGPA), passed in 1978, permitted producers to sell gas directly to end-users. Before the NGPA, producers sold gas to pipelines, which then sold to other pipelines, to LDCs, or to ultimate consumers. Today, a producer can sell directly to the consumer.

The ability to sell gas from producer, pipeline, or markets has also changed the operation of the business. Twenty years ago, two people prepared gas contracts—an engineer and a lawyer. No business decision was required, because the price and the technical aspects were governed by rules and regulations. The producer needed only to follow the technical laws and the federal and state rules.

In the wake of regulatory change, the emphasis

Figure 2. Natural Gas Movement from the Field to the Consumer— Old Process

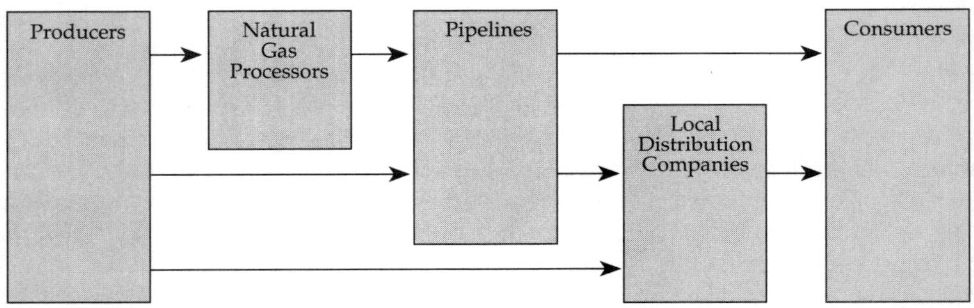

Source: Jofree Corporation.

Figure 3. Natural Gas Movement from the Field to the Consumer— New Process

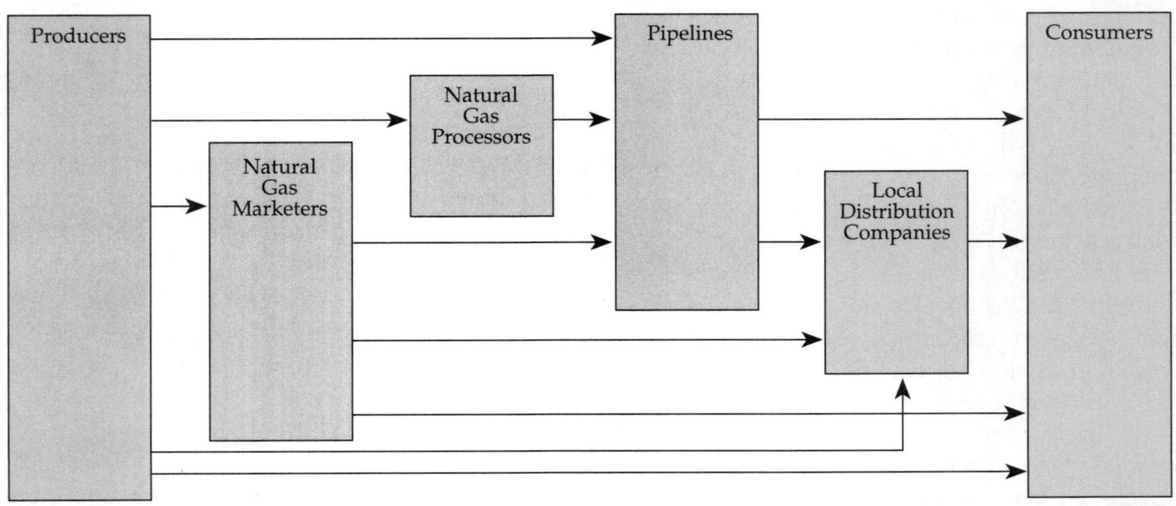

Source: Jofree Corporation.

shifted to marketing, and the industry invented the gas marketing group. **Figure 3** shows how the gas business works today. The pipeline has changed from the multifaceted role of merchant to the single role of transporter, to strictly transporting gas from one end of the pipe to the other. By 1993, when FERC (Federal Energy Regulatory Commission) Order 636 is in full operation, the pipelines' merchant activities will be reduced to almost zero.

The marketer is the new risk taker. The marketer may go so far as to take a producer's gas, find a market for it, sell it, collect the money, and pay the owner's royalty. The practice of marketers assuming such responsibilities is growing in popularity. It allows producers to concentrate their capital on the one business they know best—finding and producing gas.

Producers, pipelines, and other major sectors of the industry may own marketing companies that perform the merchant function. They should keep the two functions separate, however, because each business center needs to be a profit center on its own.

On the production side, the industry is in a maintenance mode, trying to maximize production profitability. Producers learned that marketing against competing fuels is difficult. Now living with decontrolled wellhead prices, producers must compete with other gas producers. Gas-to-gas competition has proved to be as difficult as, if not more difficult than, gas-to-oil competition. Pipelines are consolidating. Twenty-six major U.S. interstate pipelines are owned by fourteen holding companies. That number may decline to ten in the next few years. In the most recent pipeline acquisition, Koch bought United Gas Pipeline. It will probably not be a standalone company, however, because of its marketing area and the economies of scale needed for success in its area.

The pipeline transportation companies have gone to open access, which includes the unbundling of services associated with gas transportation. When producers buy transportation from the lower southwest corner of Texas to New York, for example, they no longer need to pay for storage or balancing; they can buy transportation alone.

Natural gas production is one of the few remaining venture businesses. It is a true incentive business. People in this business do not work on a rate of return. They have a venture capital mentality. Producers may sink $50 million into a well and get a dry hole, or they may make billions out of the hole. When they do hit a winner, they still have the costs for the dry holes.

Unlike producers, interstate pipelines and LDCs can be considered utilities, for which earnings are based on allowed rates of return on invested capital, not margins for value added to a product or service. Like utilities, if they have $100 million invested in pipe and equipment for a given year, and the state or federal government allows them a 15 percent rate of return, they earn $15 million from that investment for that year. They do not care about the price of the gas commodity. As long as they can keep the gas flowing or contracted to flow in the equipment in which they have invested, they earn the allowable rate of return. Therefore, until the United States adopts incentive rate making, a utility would rather spend $100 million to do a job that can be done for $10 million, because 15 percent of $100 million is more than 15 percent of $10 million.

Pricing Factors

Gas prices are a major factor in the natural gas industry. Fortunes will be made when the price is $2.60 per mmBtu, but not when it is $0.80 per mmBtu. **Figure 4** shows spot and short-term wellhead prices of gas, which averaged $1.60 in 1990, $1.38 in 1991, and $1.68 in 1992. Three conditions caused gas prices to drop in early 1992. First, the industry had too much supply and too little demand. Second, the industry has had too many marketing groups and no basis for product differentiation. If a new steel mill wants to buy gas, it can go to 300 different marketing people. The problem is they are all selling the same commodity, and the only basis on which they can

Figure 4. Average Spot and Short-Term Natural Gas Wellhead Prices, 1990–92

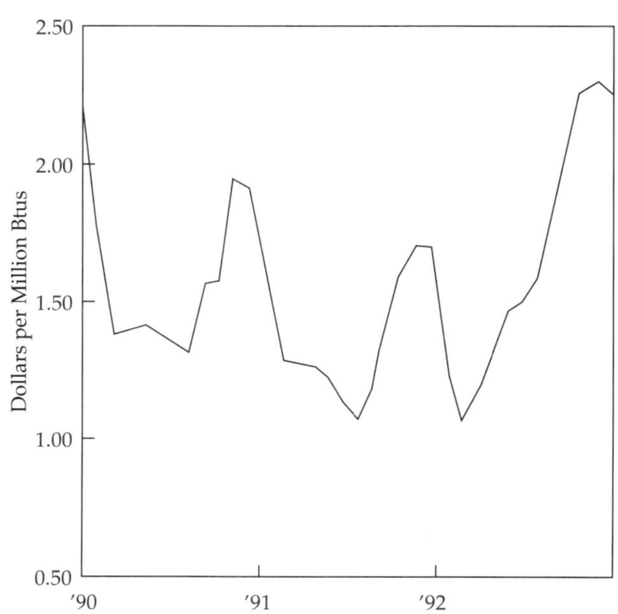

Source: Jofree Corporation.

compete is price. Third, producers moved to lower cost production through lower cost reserves. Originally, producers thought they could capture the higher margins (economic rents) for themselves. But the marketplace, including pipelines, LDCs, and consumers, recognized the cost reduction and would not pay more for the lower valued gas. Prices recovered later in 1992 from their earlier lows. This type of turnaround is typical of the gas business.

The major U.S. gas-producing areas are shown in **Figure 5**. They are the basins where gas is trapped below the surface of the earth in porous beds. The major markets are in the northeastern and north central region of the United States, in California, and in the Gulf Coast. The supply of gas is crucial and an important element in the uncertainty risk, because supply will increase at $6 per mmBtu, but at $1, producers will let reserves run down. Like any other economic commodity, supply can be enhanced by making the commodity more valuable. If it gets too costly, demand will decline because of the high cost. What was called the gas bubble will develop again.

There has been a lot of talk about the major oil companies leaving the United States. No major foreign company is currently exploring and producing gas here. In February and March 1992, when the price for newly purchased gas fell below $1, Exxon, Amoco, Mobil, and other major integrated gas companies announced that they also were moving out of the country. That decision was based on the low rates of return the producers would obtain because of the low prices. On a present-value calculation of U.S. reserves, however, these companies have more than $100 billion of remaining reserves on the books today without any new drilling. They can reduce staff, but they have too great an investment to leave completely.

Much to the companies' surprise, gas prices jumped from those lows of February. The majors are still studying the situation, however, before putting large sums back into the United States. The majors were burned in 1990, when they planned for a big 1991 and gas prices fell. Thus, they are taking a hiatus for the next four or five years until prices increase to make drilling and producing worthwhile again. Until the industry offers pricing stability, the major oil and gas companies will not drill more wells. Meanwhile, they will not leave the country because of the potential for higher prices.

Natural Gas Consumption

In 1991, 19.4 trillion cubic feet (tcf) of natural gas were consumed in the United States. **Figure 6** shows the trend in U.S. natural gas consumption. Consumption dropped in the 1970s because of lack of supply, not because of lack of market demand. The economic incentives of the period were too weak to stimulate

Figure 5. Major Gas-Producing Basins in the Continental United States

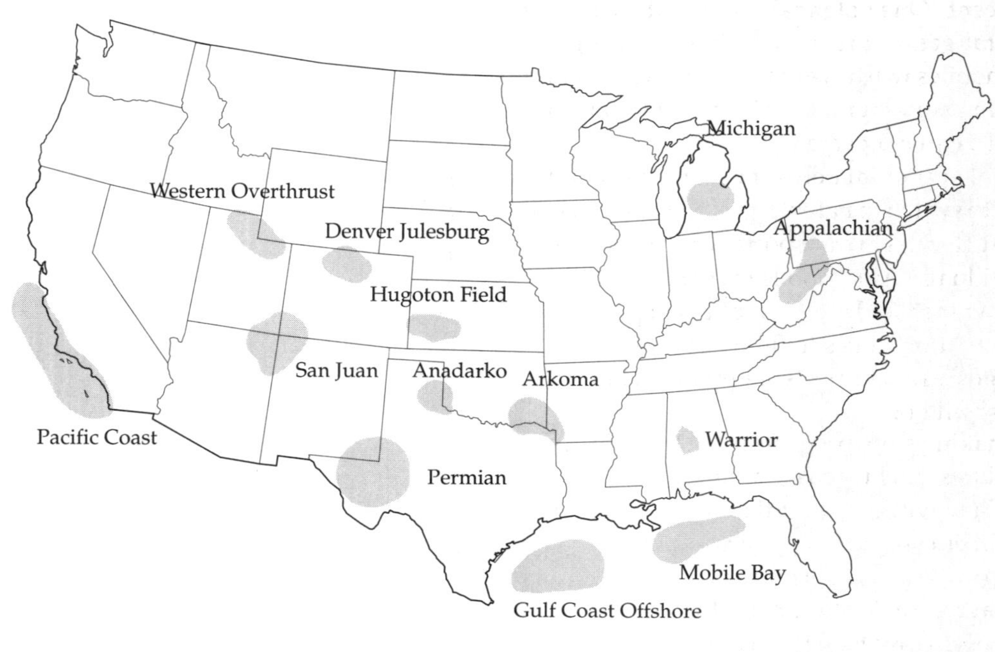

Source: Jofree Corporation.

Figure 6. U.S. Natural Gas Consumption, 1970–92

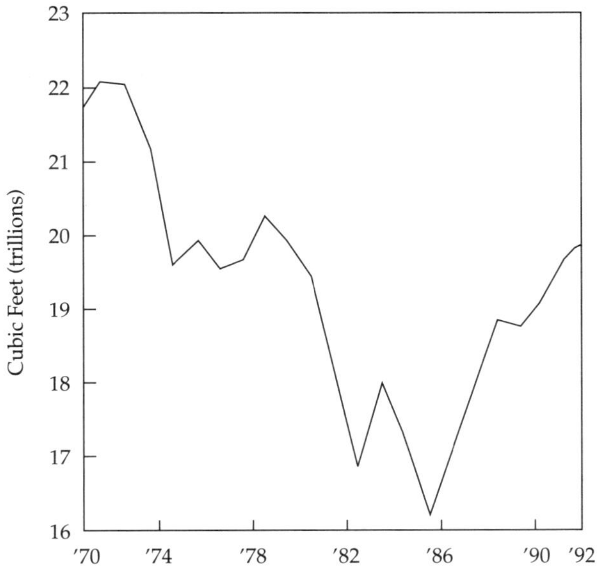

Source: Jofree Corporation.

drilling new wells. The NGPA was enacted to remedy that situation and change the industry. In the mid-1980s, high gas prices and low demand pushed consumption to its lowest levels since the early 1970s. The recent recession again helped bring all energy demand down, but consumption recovered to close to 20 tcf in 1992.

The industrial market, at 39 percent, has the largest share of the natural gas market in 1992; it is followed by residential at 24 percent, commercial at 14 percent, utility at 14 percent, and plant/pipeline fuel at 6 percent. One potential for growth lies in the use of natural gas to fuel automobiles, but a great many automobiles will have to use natural gas before that use will make a significant impact on total consumption. To equal 5 percent of today's natural gas consumption, approximately 4 million automobiles would have to switch to natural gas. Natural gas will become useful as fuel in transportation, but probably not by individuals. Automobile owners are unlikely to get out a wrench and a pair of pliers to plug a gas line into their automobiles to fill up overnight. Corporate and government fleets might do it, but ordinary drivers will not.

Gas is making slow progress into electricity generation. Utilities can buy coal on the basis of 20-year contracts. They will pay an inflation factor, but the coal price will be essentially flat. Prices of oil and gas, as seen in **Figure 7,** are much more volatile. Until gas companies agree to index prices to coal or some stable index, the growth of gas in the utility industry will be limited.

Political Changes

Currently, the big question mark is FERC Order 636. Martin Allday, the current FERC chairman, has been the standard-bearer for Order 636. If he leaves the commission, some faults of the order will come under review. The industry has gone too far, however, in shifting control to the marketing sector and making pipelines transporters only to revert back to the old pattern. Regulation never leads; it always follows. Congress and the regulators are not fearless enough to be leaders; they are always reactive. The marketplace is what leads, certainly in the gas industry. The marketing element is what holds the business together, and that will not change.

The Clinton administration will not make any radical changes in the natural gas business. Although President Bill Clinton promised to push the use of natural gas, the new administration will do little for the gas industry. Vice President Al Gore will strongly support the use of gas in boilers and other facilities, but his environmental viewpoint could cause problems for gas companies in the drilling and exploration areas; the Clinton administration will do little to open various areas where oil and gas exploration could be beneficial. Personnel and policy changes at FERC and the Department of Energy are likely, but on a net-benefit basis, the government will have little effect on the industry. The marketplace

Figure 7. Prices of Various Fuels Delivered to Electric Utilities, 1973–91

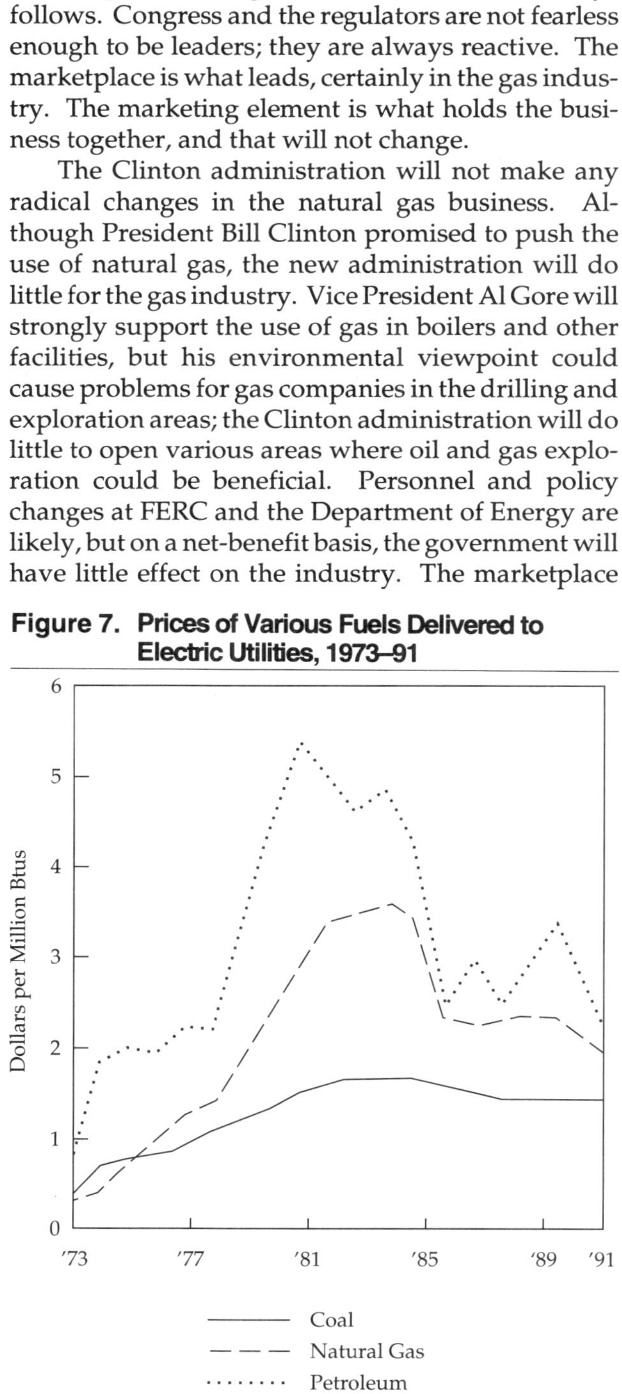

Source: Jofree Corporation.

rules, however, so the expanded use of natural gas will probably bring changes in the long term.

Summary and Future Prospects

On balance, the natural gas business is now a stand-alone industry, no longer the stepchild of black oil. The gas business is its own profit center. Of course, economics dictate: Decisions on drilling a well are based on the gas it might produce. Producers will not drill a well just because the location is a nice geological formation. Drilling must be a good engineering move; it must be financially understood and probable; and if it is successful, the gas must be close to transportation and provide a suitable rate of return.

Over the next three years, overproduction and oversupply of natural gas should balance out. Many marketing companies will vanish. Producers will learn that if they produce higher priced gas, they will receive higher prices, and if they produce lower priced gas, they will receive lower prices. Gas is the fuel of the 21st century. It is a domestically available fuel, it is an abundant fuel, and it is an environmentally acceptable fuel. Demand will not, however, jump to 25 or 30 tcf overnight, nor will prices increase sharply. Reasonable average wellhead prices would be $1.66 per mmBtu for 1992, $1.72 for 1993, $1.80 for 1994, and $2 by 1995. Oil prices should be in the $20-a-barrel range.

The trend is toward orderly expansion of the gas business. (Of course, no one can predict hurricanes or wars.) The future of the industry all comes down to prices; regulation is secondary. Stability in prices will increase demand. The key element, then, is getting rid of the uncertainty.

Question and Answer Session

Carol Freedenthal

Question: How does the Clean Air Act affect the fuel-switching decision? Will natural gas become a premium fuel for utilities' boiler applications?

Freedenthal: The Clean Air Act must always be taken with a grain of salt. If the economics are there, the act will drive people to use natural gas. If gas prices respond by rising too sharply, the use of cleaned-up coal or oil will rise. Eventually, utilities will go to nuclear power.

Question: Will Section 29 tax credits be resurrected?

Freedenthal: Section 29 refers to the credits given for tight sands gas, coal methane gas, biomass, and fuel conversion. Although it was scheduled to expire several years ago, Congress extended the credits several times. In 1992, Congress did not extend the drilling deadline under Section 29. Thus, producers that start active drilling programs for wells after December 31, 1992, cannot receive tax credits for those wells. As a result, producers were drilling heavily in late 1992 to start additional wells before year-end. For wells active before the deadline, the tax incentive runs through 2001. It is a good tax incentive for producers who can take advantage of it. Ending the credit is a good move for the industry, however, because it was such a great incentive that producers were tempted to drill more than they needed. I doubt the credit will be resurrected under the Clinton administration.

Question: What was the reason behind this provision?

Freedenthal: President Jimmy Carter said that the lack of gas is almost like an act of war and we need to boost gas production. Congress enacted these various incentives to increase the supply of gas. What they found was that the best incentive to get producers to drill wells is to raise the price. There is no substitute for good economics.

Valuing Natural Gas Pipeline Securities

Richard G. Gross II
First Vice President
Lehman Brothers

Analysts must examine both sides of the funding process—equity and debt—when valuing natural gas pipeline stocks.

Factors impinging on the industry, changes within the industry, and rate making—all have influenced the stock prices of pipeline securities. Much of the industry's evolution is related to the changing nature of pipeline regulation. To predict future performance, however, investors must look at the whole valuation process, which is materially different for pipelines than for most other investments.

Industry Characteristics and Performance

The pipeline industry has been in a state of flux for 20 years. In the 1970s, the natural gas industry suffered through government-induced supply shortages, an enormous rise in commodity prices, and cooler-than-normal weather. In the 1980s, it endured chronic oversupply of gas, massive consolidation, sweeping regulatory change, a price collapse followed by enormous volatility, and a demand collapse followed by a partial recovery.

Many problems will be alleviated in the 1990s as the industry implements a new rate design and the outlook for demand continues to be positive. Gas prices will remain volatile but with an upward bias.

External influences on the gas industry have varied so dramatically during the past 20 years that a normalized valuation period cannot be established. Industry returns relative to the S&P 400 have been extremely volatile, as **Figure 1** shows, and much of this volatility is related to the macro backdrop—the external environment in which the industry has operated. Regulatory change (see **Figure 2**), weather (see **Figure 3**), and gas prices (see **Figure 4**) are important influences that have had varying degrees of impact on the stock price performance of the pipeline companies. Gas prices, for example, have a stronger influence on the fortunes of the independents than on those of the pipelines, at least in cold weather. (Note that Figures 2, 3, and 4 present the stock performance for an index of nine companies relative to the performance of the S&P 400. The nine-company index allows for continuity in industry performance and adjustments for write-offs. The index is composed of Arkla, Coastal Corporation, Consolidated Natural Gas Company, Panhandle Eastern Corporation, Sonat, Transco Energy Company, ENSERCH Corporation, Equitable Resources, and the Williams Companies.)

The relative performance of the local distribution companies (LDCs) closely tracks the inverted 30-year T-bond yield, as shown in **Figure 5.** Thus, LDC stocks can be used to some degree as bond substitutes.

The average actual return on equity for pipelines, as demonstrated by the nine-company index shown in **Figure 6**, has fallen below the average allowed return—that is, the regulated return—since the mid-1980s. Prior to 1984, companies earned or overearned the allowed rate of return, which rose when interest rates rose.

The average book value per share for pipeline companies has dropped in recent years. **Figure 7** shows that the nine-company average declined from 1982 to 1992. The average dropped from a high of $25 a share in 1984 to a low of $17 a share in 1988 before recovering somewhat after 1989.

Company Differences

Pipeline companies are not a homogeneous group. The companies differ widely in asset mix, in financial stability and strategy, and in the methods used to account for exploration, production, and acquisition activities.

Figure 8 illustrates the large differences in asset composition among 10 major pipeline companies. It shows the amount of assets the companies allocated

Figure 1. Volatility of Pipeline Company Returns Relative to the S&P 400, 1969–3Q'92

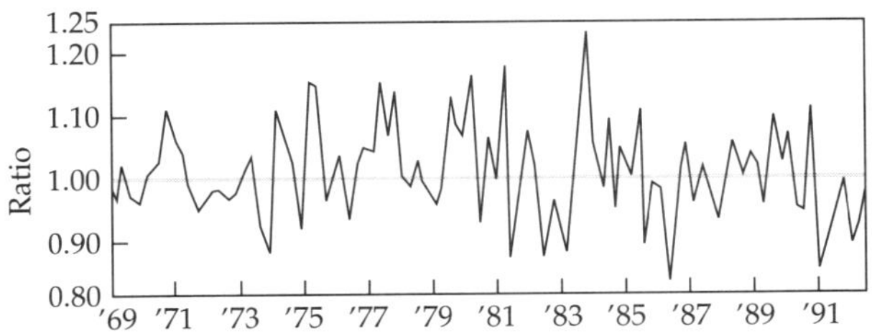

Source: Lehman Brothers.

Figure 2. Influence of Regulation on Pipeline Stock Relative Performance, 1983–3Q'92
(nine-company index relative to S&P 400)

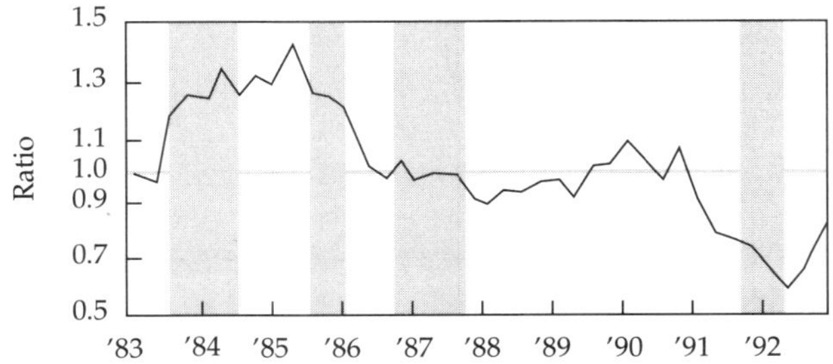

Shaded areas represent periods of changes in federal regulation.

Source: Lehman Brothers.

Note: Nine-company index performance is very close to broader Moody's and S&P indexes.

in 1991 to gas distribution, gas transmission, exploration and production (E&P), other energy activities, and nonenergy activities. Different asset mixes can lead to large differences in the nature and variability of the companies' cash streams. Many of the assets yield unpredictable cash flow streams because they are exposed to natural gas prices and weather conditions.

Cash flow—what companies use, deploy, and reinvest—is materially different from the earnings that appear in financial statements. Cash flow of LDCs is sensitive to weather fluctuations; E&P companies' cash flows are sensitive to changes in natural gas prices. Both are unpredictable.

The volatility of cash flows for a company depends partly on the company's rate design. If a company adopts the straight fixed/variable rate rather than the modified fixed/variable rate, the volatility of the company's cash flow will decline, but the effect is uneven within the industry. Pipeline companies operating under volumetric rate designs share the same sensitivity to weather as the LDCs. New regulation (Order 636) will change the volatility of earnings and cash flows.

Accounting policies among natural gas companies vary widely. Therefore, analysts must disaggregate and then reassemble reported earnings to make valid comparisons. E&P companies can choose between two different types of accounting, one more liberal than the other. The full-cost method, by capitalizing unproductive well costs, overstates earnings in relation to the successful-efforts method, which expenses such costs.

Different financial strategies translate into different levels of financial stability for these companies. Some managers like to run the companies at the very edge of investment grade; others prefer an Aa bond rating because it has a significant impact on the valuation of their securities. After adjustment for working capital positions, debt-to-capitalization ratios range from about 40 percent to more than 70

Figure 3. Influence of Weather on Pipeline Stock Relative Performance, 1983–3Q'92
(nine-company index relative to S&P 400)

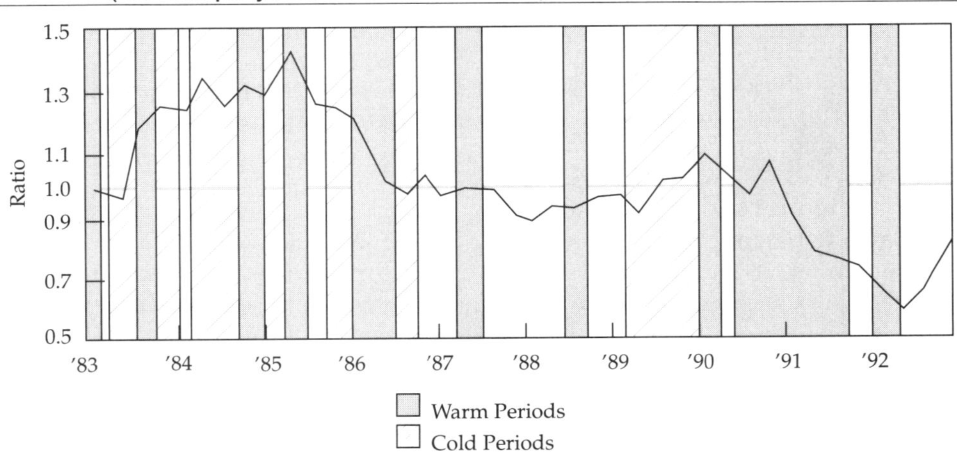

Source: Lehman Brothers.

Figure 4. Influence of Gas Prices on Pipeline Stock Relative Performance, 1983–3Q'92
(nine-company index relative to S&P 400)

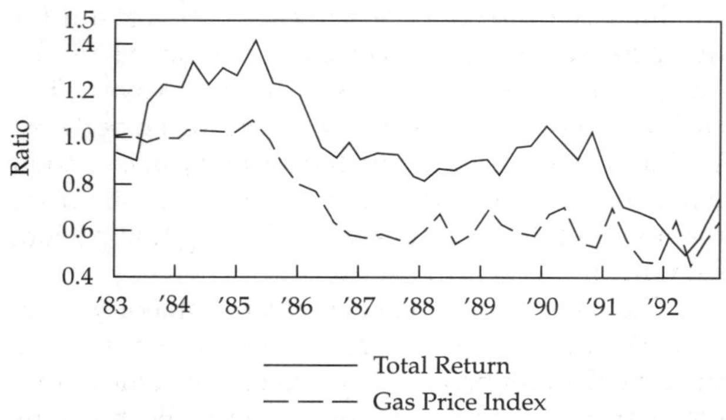

Source: Lehman Brothers.

Figure 5. LDC Relative Stock Performance and 30-Year T-Bond Yields, 1991–92

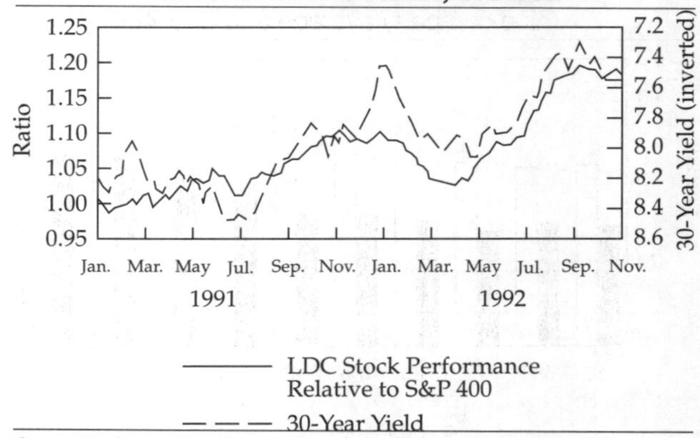

Source: Lehman Brothers.

Figure 6. Actual and Allowed Returns on Equity, Interstate Natural Gas Pipelines, 1982–91

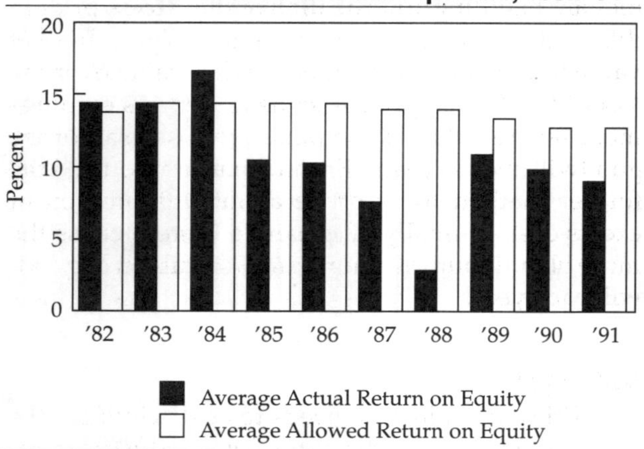

Source: Interstate Natural Gas Association of America.

Figure 7. Book Value per Share, 1982–91
(nine-company average)

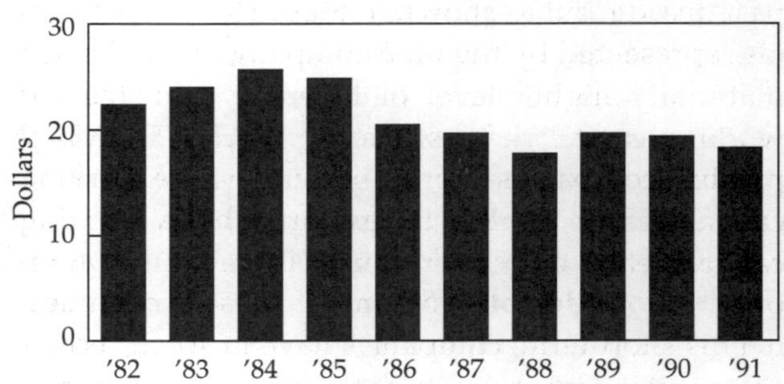

Source: Lehman Brothers.

Note: Nine-company index performance is very close to broader Moody's and S&P indexes.

Figure 8. Composition of Assets—10 Major Pipeline Companies, 1991

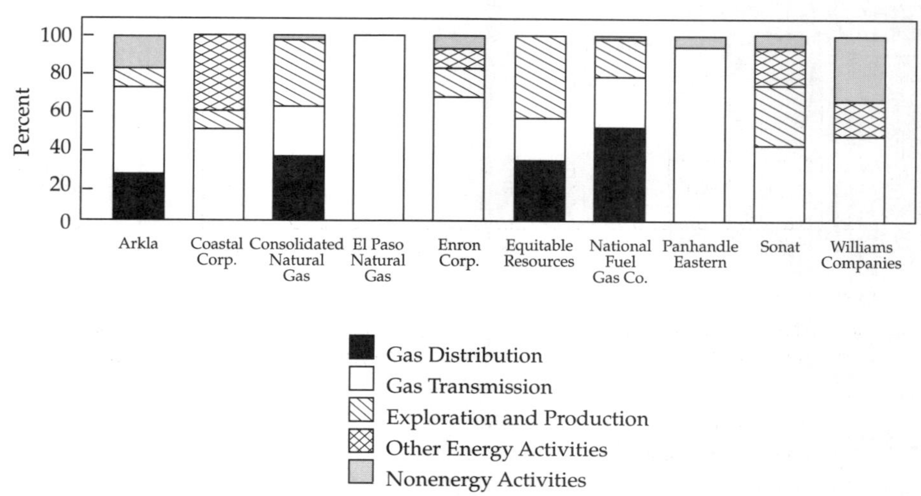

Source: Lehman Brothers.

Figure 9. Ratio of Adjusted Debt to Adjusted Capitalization, Year-End 1982–91
(nine-company average and S&P 400)

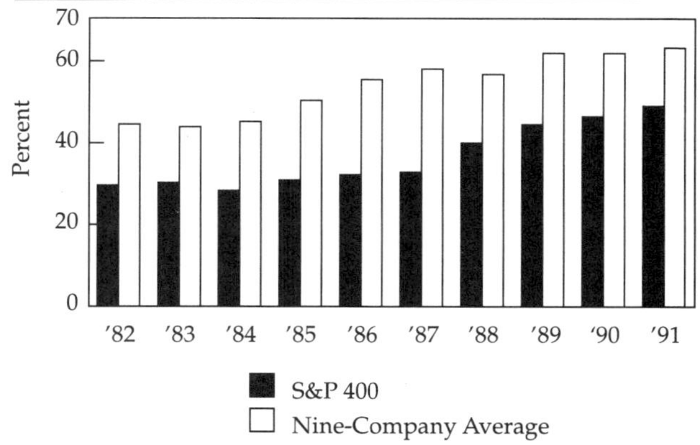

Source: Lehman Brothers.

percent.

Both the S&P 400 companies and the pipeline companies have leveraged their balance sheets in the past decade, but as shown in **Figure 9**, the pipelines (as represented by the nine-company index) have a materially higher level of leverage, adjusted for working capital deficits, than the S&P 400. Many pipeline companies operate under massive working capital deficits. Arkla, for example, has a working capital deficit of greater than $900 million. Moreover, a considerable amount of these companies' debt is short term; companies have factored receivables and incurred significant liabilities for customer refunds.

The companies also have varying levels of capital intensity. Independent production operations are very capital intensive, and regulated utilities generate massive amounts of predividend cash flows.

Increased takeover activity was one reason for the increased leverage. As shown in **Figure 10**, the industry has experienced two takeover waves. The early and more active wave, which occurred between 1983 and 1986, came before the industry understood what the major changes in the regulatory environment would do to the industry. During that period, four major mergers occurred.

The second wave was actually one major merger in 1989 and one departure in 1991, when a major pipeline filed for bankruptcy protection (a move that some argued was regulation driven). Both events had a major impact on how pipeline stocks traded in the marketplace and on their relative values.

The purchase prices in the takeovers generally exceeded book values, and the purchasing companies are handling amortization of this excess price in different ways. One method is to assign a market value to the assets and amortize that value over the asset life. This approach tends to depress earnings more severely than if companies choose to amortize goodwill over 40 years. Enron, for example, used the former method to amortize about $100 million of excess costs annually. Panhandle Eastern chose the more liberal route of amortizing $15 million of goodwill each year.

Valuation

Although valuation needs to be disciplined, the macro influences on the natural gas industry intro-

Figure 10. Takeover Activity, 1982–91

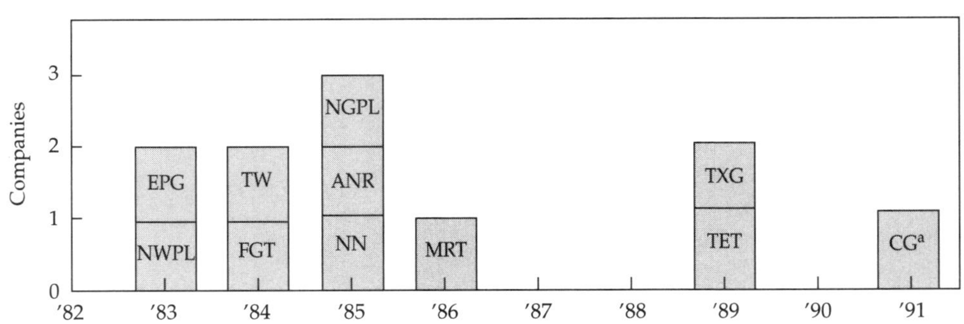

[a]Bankruptcy.

Key:

ANR	= ANR Corp.	NN	= Northern Natural Gas
CG	= Columbia Gas System	NWPL	= Northwest Pipeline
EPG	= El Paso Natural Gas	TET	= Texas Eastern Corp.
FGT	= Florida Gas Transmission	TW	= Transwestern Pipeline
MRT	= Mississippi River Transmission	TXG	= Texas Gas Transmission
NGPL	= NG Pipeline		

Source: Lehman Brothers.

Figure 11. Historical Measures of Pipeline Companies' Relative Value, 1982–92
(relative to the S&P 400)

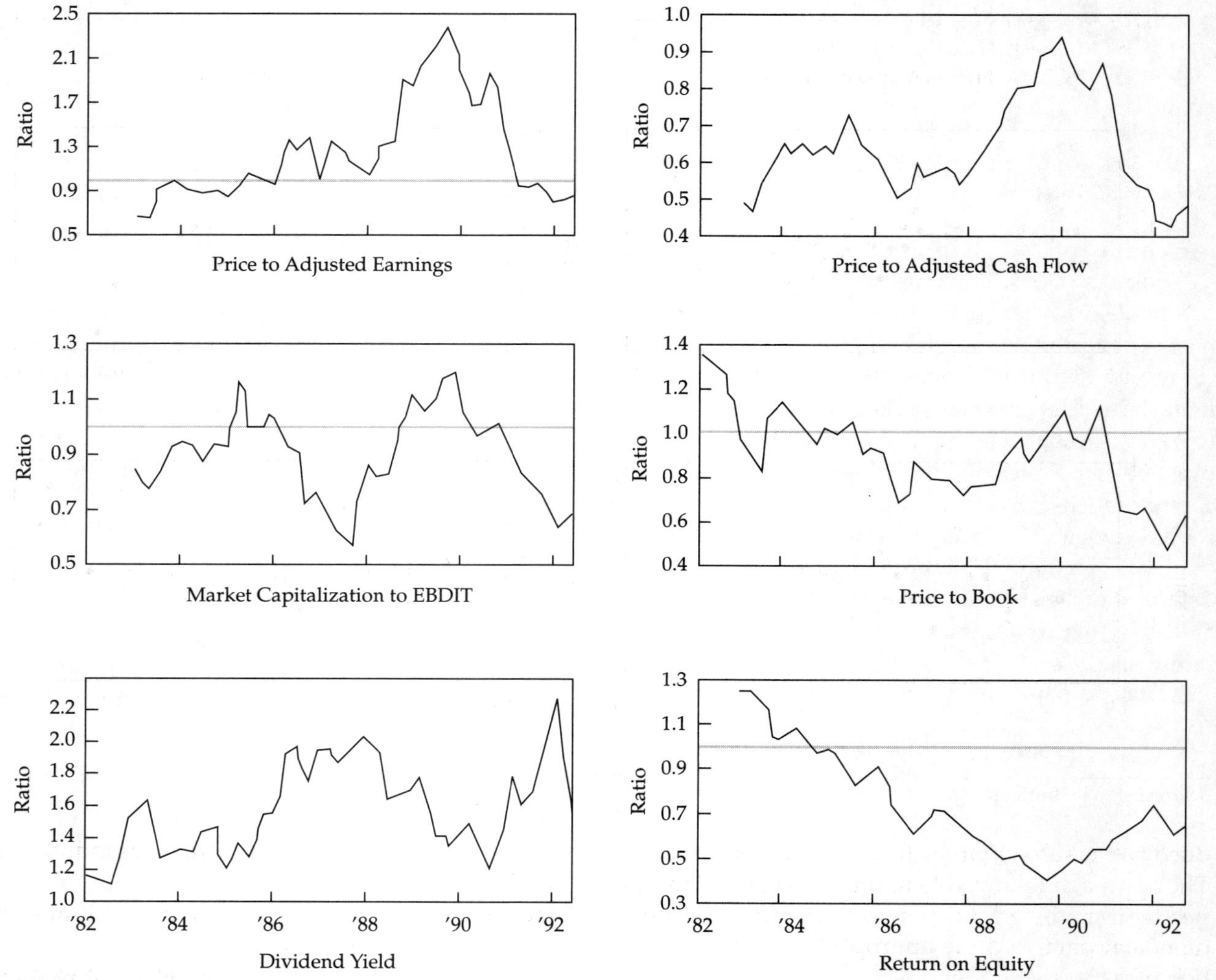

Source: Lehman Brothers.

Figure 12. Measures of Relative Value as Predictors of Performance (relative to the S&P 400)

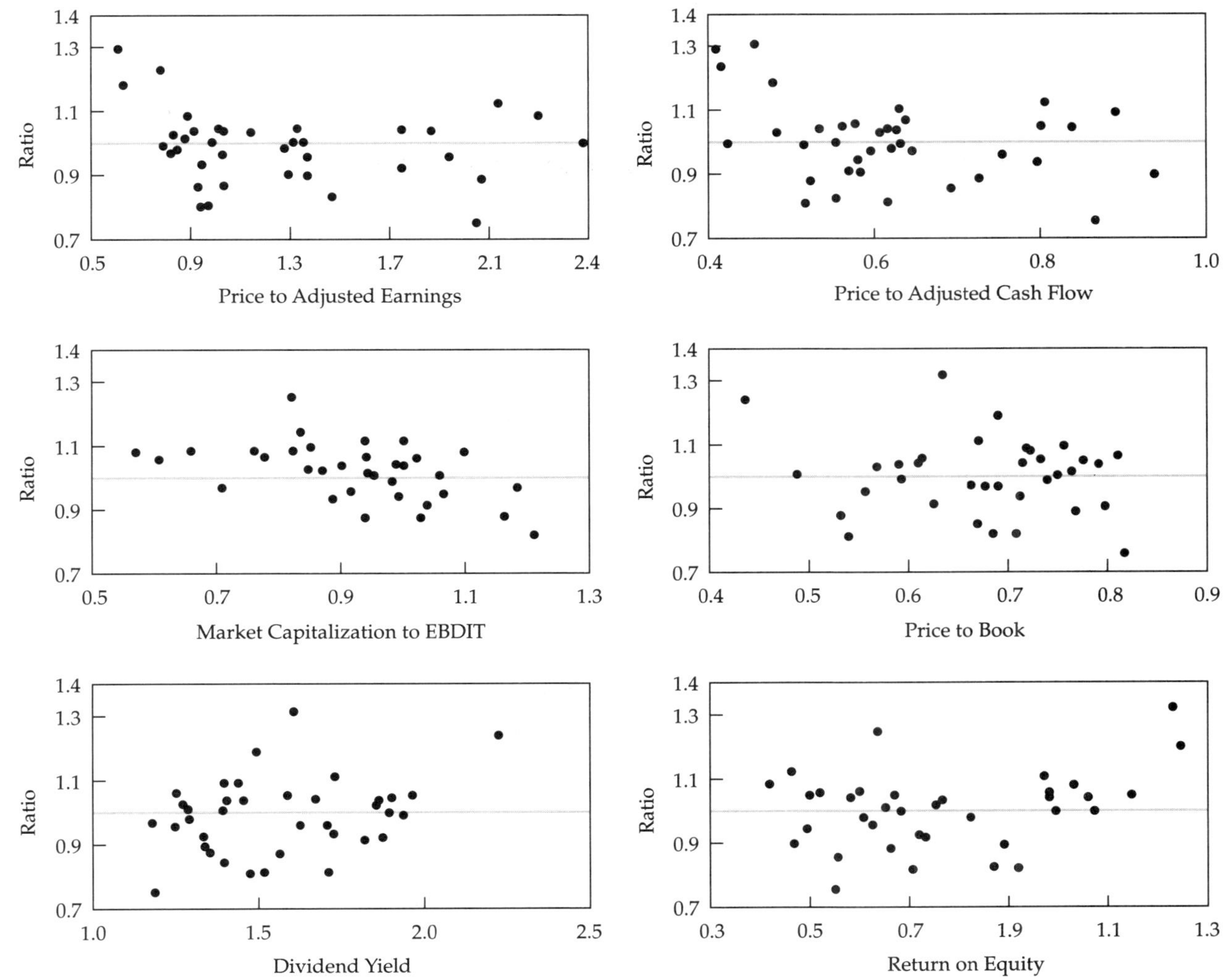

Source: Lehman Brothers.

duce some subjectivity into the valuation process. The approach followed here thus considers four aspects: macro factors, micro factors, comparative fundamental data, and the appropriate relative valuation measures.

Macro factors. Because external factors affect what the market will pay for pipeline earnings or cash flow streams, analysts must value the stocks in the context of the key macro factors: regulation, energy prices, weather, the state of the economy, and the general level of the financial markets.

Micro factors. Key micro factors—those that are company specific—are asset mix, accounting policies, condition of the balance sheet, and management. For example, the market has embedded a premium in the stocks of companies that were well managed through the recent regulatory turmoil in the pipeline industry. In the natural gas industry, in which operations are strongly influenced by factors outside management control, management must be proactive instead of reactive.

Comparative fundamental data. To examine fundamentals, analysts should compare returns, coverage ratios, capital ratios, and growth.

Valuation measures. Analysts typically use the valuation measures shown in **Figure 11** to measure historical performance. The key valuation measures for the pipeline companies are relative price to earnings, relative price to cash flow, and market capitalization adjusted by EBDIT (earnings before depreciation, interest, and taxes). Figure 11 gives actual data for the nine-company index. Of these measures, only two have proved to be good predictors of superior stock performance: extremely low ratios of relative price to earnings and price to cash flow. The latter is typically experienced after crises and has occurred twice in recent history—before oil prices cratered in 1986 and before the promulgation of Order 636, when people feared more take-or-pay charges. The scatter diagrams of relative value as a predictor of near term performance for the nine-company index appear in **Figure 12**. Note the randomness; none of the scatterplots exhibits a strong correlation between relative performance and relative valuation.

Regression analysis helps determine which factors work best when valuing these companies. As shown in **Figure 13**, the "best fit" line for the pipeline index, gas prices, book value, regulation, weather, and year-to-year changes in the economy have worked reasonably well as predictors of past performance of pipeline stocks. The first three factors have been by far the most important, the latter two much less so.

Table 1, **Table 2**, **Table 3**, and **Table 4** provide relative return ratios, capital ratios, coverage ratios, and valuation measures, respectively, for 11 companies. The ratios compare 1991 actual values with 1992 and 1993 estimated values. Using variations of measures in each category helps clarify the differences among the companies.

Growth prospects are important. If demand grows 2 or 3 percent a year, rate-base growth will not be high; rate-base growth comes from changes in supply, not demand. Based on current pipe configuration, about 6 of 20 pipelines will have rate-base growth in excess of 5 percent.

Figure 13. Computed and Actual Cumulative Total Pipeline Company Returns, 1983–2Q'92
(nine-company index relative to the S&P 400)

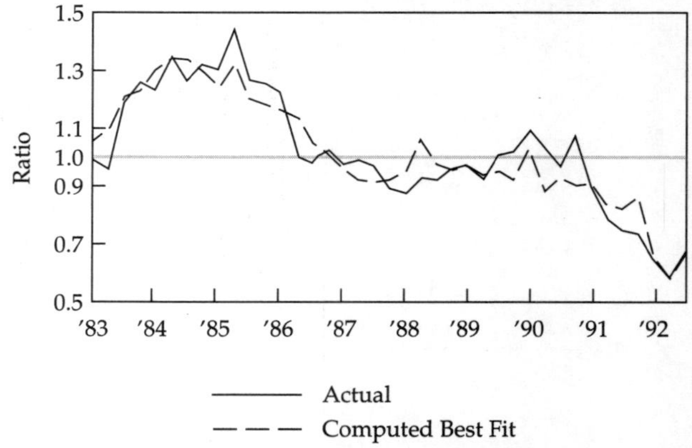

Source: Lehman Brothers.

Arkla illustrates the importance of using several valuation measures instead of relying on one. Based on Arkla's estimated P/E for 1993 (see Table 4), the company's stock is selling at less than a market multiple. Based on cash flow, the stock is dirt cheap; anyone can make money buying stocks at three times cash flow. A cross-check that includes Arkla's debt in the valuation, however, will reveal that Arkla is one of the most expensive stocks in the industry.

In summary, analysts must look at both sides of the funding process—equity and debt—when valuing these stocks. Remember that pipeline companies are highly leveraged, so large portions of their earnings are being used to pay back the banks.

Table 1. Comparison of Returns, 11 Pipeline Companies

	Return on Average Equity			Cash Flow from Operations (CFFO)/Adjusted Capital			EBDIT/Adjusted Capital		
Company	1991	1992[a]	1993[a]	1991	1992[a]	1993[a]	1991	1992[a]	1993[a]
Arkla	1.9%	4.6%	7.5%	8.7%	8.0%	8.6%	12.9%	12.8%	13.6%
Coastal Corp.	4.8	7.6	10.8	6.1	8.2	9.3	15.7	17.2	19.3
Consolidated Natural Gas Co.	9.0	11.1	12.1	14.6	15.3	16.3	18.1	19.8	20.9
El Paso Natural Gas	10.8	10.1	11.9	3.2	6.1	12.4	21.6	18.3	18.6
Enron Corp.	13.1	14.4	13.9	12.9	15.2	16.6	22.2	24.0	25.4
Equitable Resources	11.8	10.8	13.4	12.7	13.1	16.2	19.2	19.8	23.6
National Fuel Gas Co.	9.5	9.7	10.4	10.8	9.3	9.3	13.6	14.2	14.7
Panhandle Eastern Corp.	6.9	7.5	8.7	8.3	9.3	9.7	17.6	17.1	17.2
Sonat	7.4	8.1	9.9	13.8	14.9	17.5	20.1	21.2	23.5
Transco Energy Co.	3.0	5.2	13.1	9.1	13.8	16.9	16.2	20.9	25.0
Williams Companies	9.4	9.5	10.3	10.4	11.5	12.0	16.8	15.7	17.1
Average	8.0%	9.0%	11.1%	10.1%	11.3%	13.2%	17.6%	18.3%	19.9%

Source: Lehman Brothers.

[a]Estimate.

Table 2. Comparison of Capital Ratios, 11 Pipeline Companies

	Total Debt/Total Capital			Adjusted Debt/Adjusted Capital			Common Dividend Payout Ratio		
Company	1991	1992[a]	1993[a]	1991	1992[a]	1993[a]	1991	1992[a]	1993[a]
Arkla	73.0%	73.7%	69.7%	74.3%	75.6%	72.4%	NM	179.2%	55.9%
Coastal Corp.	69.4	68.8	67.0	65.1	64.3	62.1	43.7%	30.3	21.6
Consolidated Natural Gas Co.	46.1	47.3	47.8	39.8	40.7	40.6	96.7	78.8	72.5
El Paso Natural Gas	50.3	46.4	47.8	44.5	45.3	45.8	0.0	47.7	48.0
Enron Corp.	65.5	53.0	48.9	65.3	54.5	50.0	57.7	53.7	46.9
Equitable Resources	46.3	46.3	44.0	42.8	41.8	38.9	48.7	52.3	41.7
National Fuel Gas Co.	55.0	55.9	56.5	53.5	51.5	52.3	87.6	87.7	86.0
Panhandle Eastern Corp.	65.2	64.6	63.5	66.2	65.8	64.8	93.0	85.7	72.4
Sonat	57.2	51.2	49.3	52.8	47.3	45.8	110.1	95.2	74.1
Transco Energy Co.	86.5	87.7	82.8	87.7	88.6	83.8	234.4	95.8	43.1
Williams Companies	61.8	59.3	60.9	64.3	61.9	63.0	59.6	61.8	55.2
Average	61.5%	59.5%	58.0%	59.7%	57.9%	56.3%	77.5%	80.0%	56.1%

Source: Lehman Brothers.

Note: NM = not a meaningful figure.

[a]Estimate.

Table 3. Comparison of Coverage Ratios, 11 Pipeline Companies

	EBDIT/Interest and Preferred Dividend			CFFO/Total Debt			Net CFFO/Capital Expenditures		
Company	1991	1992[a]	1993[a]	1991	1992[a]	1993[a]	1991	1992[a]	1993[a]
Arkla	2.2	2.1	2.3	8.7%	8.0%	8.6%	50.2%	110.6%	145.1%
Coastal Corp.	2.0	2.2	2.5	7.7	10.6	12.3	43.0	74.6	87.2
Consolidated Natural Gas Co.	5.1	5.3	5.6	28.4	28.8	29.4	61.3	77.6	91.1
El Paso Natural Gas	2.5	3.7	3.7	38.3	38.0	36.8	17.2	23.6	78.2
Enron Corp.	2.9	3.1	3.5	19.6	29.7	34.6	62.3	114.3	98.1
Equitable Resources	5.7	5.3	6.7	28.8	28.5	36.7	39.5	91.0	108.4
National Fuel Gas Co.	2.5	2.8	3.1	19.0	16.5	16.3	49.7	49.9	58.3
Panhandle Eastern Corp.	2.1	2.3	2.4	13.5	15.1	16.2	107.2	109.6	121.4
Sonat	3.3	3.5	4.2	21.6	26.6	30.8	45.7	94.5	116.3
Transco Energy Co.	1.7	2.0	2.4	21.2	26.5	31.9	38.8	121.6	154.1
Williams Companies	2.8	2.8	2.9	18.4	22.3	22.4	71.6	41.1	63.5
Average	3.0	3.2	3.6	20.5%	22.8%	25.1%	53.3%	82.6%	102.0%

Source: Lehman Brothers.

[a]Estimate.

Table 4. Comparison of Valuations, 11 Pipeline Companies

	Price/Earnings			Price/Cash Flow			EBDIT/Adjusted Market Capitalization		
Company	1991	1992[a]	1993[a]	1991	1992[a]	1993[a]	1991	1992[a]	1993[a]
Arkla	95.8	28.8	15.7	3.3	3.7	3.4	8.0	8.2	7.6
Coastal Corp.	27.7	17.0	11.1	6.3	5.4	4.6	6.8	6.1	5.3
Consolidated Natural Gas Co.	23.6	21.8	17.0	9.8	8.0	7.2	8.7	8.1	7.2
El Paso Natural Gas	14.7	13.8	12.6	32.6	13.2	5.9	6.3	6.2	5.7
Enron Corp.	23.6	20.5	17.2	7.3	6.3	5.6	6.9	6.3	5.9
Equitable Resources	15.3	15.8	12.0	7.1	6.8	5.4	7.5	7.0	5.7
National Fuel Gas Co.	16.7	15.6	13.6	6.5	6.1	5.8	7.7	7.4	7.0
Panhandle Eastern Corp.	21.4	19.3	16.7	5.5	5.3	5.1	6.1	6.9	6.9
Sonat	24.1	22.5	16.3	6.2	5.8	4.9	6.5	6.2	5.6
Transco Energy Co.	NM	22.1	10.3	2.2	1.4	1.6	5.3	5.2	5.5
Williams Companies	16.1	15.5	13.1	5.4	4.8	4.3	6.9	7.1	6.2
Median	22.5	19.3	13.6	6.3	5.8	5.1	6.9	6.9	5.9
Average	27.9	19.3	14.1	8.4	6.1	4.9	7.0	6.8	6.2
S&P 400	29.2	21.2	17.5	11.0	9.0	8.4	8.1	6.7	6.2
Pipeline average/S&P 400	95.7%	91.3%	80.7%	76.4%	67.4%	58.2%	86.0%	101.9%	100.2%

Source: Lehman Brothers.

Note: NM = not a meaningful figure.

[a]Estimate.

Valuing Natural Gas Securities—Local Distribution Companies

Michael T. Kerr
Vice President
Capital Guardian Research Company

Although local distribution of natural gas might appear on the surface to be a simple business, it is in fact complex. Politics, for example, plays a major role in valuation because of conflicting interests among producers, pipelines, and consumers; thus, analysts have the task of predicting shifting coalitions.

The natural gas industry, particularly the local distribution company (LDC) segment, is fun to analyze. For one thing, LDCs have outperformed the market while the rest of the industry languished: In 1985, many pipeline and exploration company stocks were selling in the $20s and $30s; today, they all are selling in the teens.

This presentation reviews the effects of the evolution of the natural gas industry on LDCs and discusses how to value LDCs. What seems to be such a simple industry is, in fact, becoming complex.

Evolution of the Industry

The regulated part of the natural gas business is decreasing in importance, and the value added by good management is growing. **Figure 1** and **Figure 2** illustrate the evolution of the gas supply system. Figure 1 shows the basic movement of gas through the various segments of the industry and the prices charged in each segment in 1991. Under this system, public utility commissions monitored only one official price transaction, that between LDCs and end-users. In late 1993, when Federal Energy Regulatory Commission (FERC) Order 636 takes effect, it will continue the unbundling of pipeline services that has been occurring and complicate the distribution system even more. Under Order 636, LDCs will have a much more complex supply environment than previously, and as depicted in Figure 2, public utility commissions will have to monitor six possible transactions instead of one. The increase in transactions raises the possibility of disallowed LDC gas purchases.

The LDC share of natural gas sales is declining, as **Figure 3** shows. Even the recent growth in natural gas demand has not bolstered LDC sales, because much of the growth in demand has been fueled by mainline sales from the pipeline companies directly to industrial and electric utility customers. Moreover, this bypassing trend will probably continue for the next 5–10 years. Projections of natural gas demand from 1990 to 2005 show the bulk of growth occurring in places where LDCs have a competitive disadvantage rather than an advantage.

LDCs collected most of the economic rent in the industry in the 1980s. **Table 1** shows a history of natural gas prices and margins at the wellhead and in residential and industrial markets from 1967 through 1990. In general, producers have had difficulty maintaining prices, and interstate pipeline margins, excluding take-or-pay charges, have flattened out.

Figure 4 shows the wellhead price as a percentage of residential, commercial, and industrial prices. As LDCs fought to maintain industrial market share, much of the cost recovery was shifted to commercial and industrial customers. Thus, the wellhead price became a much smaller portion of these customers' prices than it was previously. This cost shift is illustrated in **Figure 5** by the LDCs' increasing percentage of total residential gas price. Although the wellhead price dropped 50 percent from 1984 to 1991, the residential price remained flat, at $6, because the LDCs were recovering more costs from this market.

Valuation Issues

Politics will always be part of valuation in the

Figure 1. Natural Gas Chain of Supply, 1991

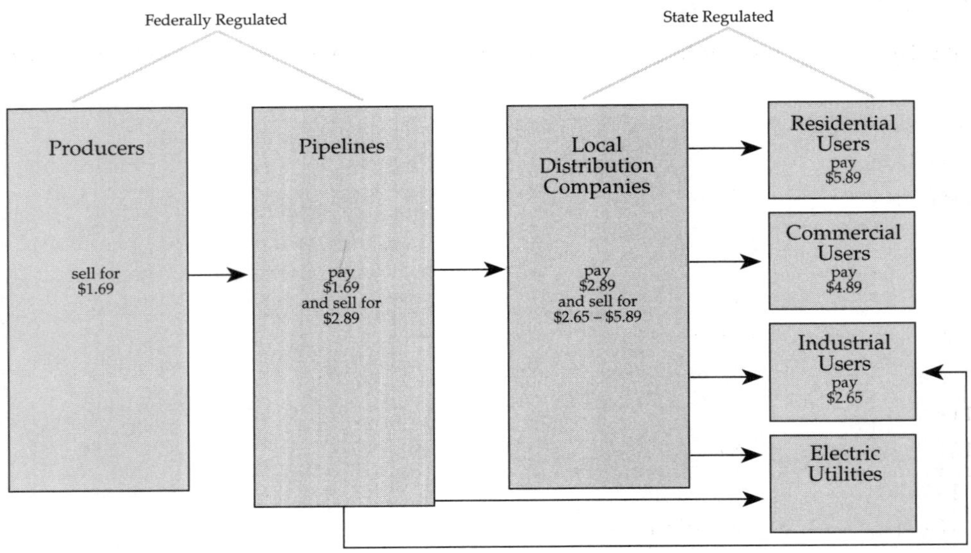

Source: Capital Guardian Research Co.

Note: The prices charged at each segment are based on American Gas Association data.

Figure 2. Transaction Paths for the Natural Gas Industry under Order 636

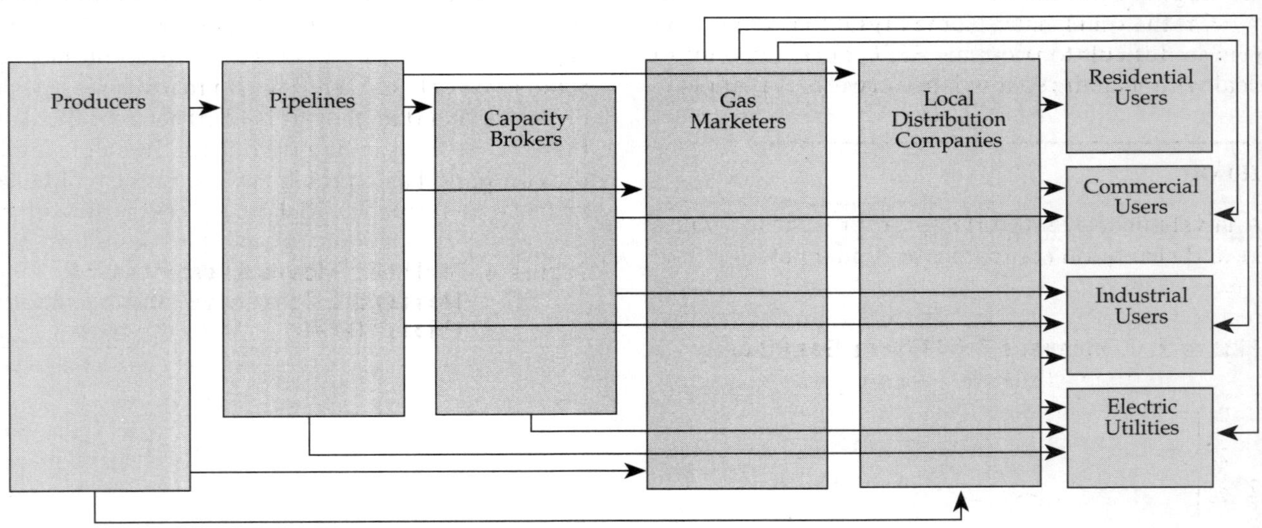

Source: Capital Guardian Research Co.

natural gas industry because of conflicting interest groups. The industry has three basic constituencies—the producer, the pipeline in the middle, and the consumer (represented by the LDC). Generally, in regulatory affairs, two constituents gang up on the third one. During the 1985–87 period, LDCs were in most cases part of the winning coalition. Together with producers, the LDCs thrashed the pipelines. In 1992, however, the pipelines and producers realized they had a common goal—to pass along take-or-pay and transaction costs—and alliances shifted.

Analysts need to predict the next coalition. On the one hand, some believe that Order 636 was the result of a FERC majority representing producers' interests. They see Order 636 as a way for producers to trounce the LDCs and consumers. Order 636 allows more pipeline costs to be passed directly through to residential customers, who will thus bear

Table 1. History of Natural Gas Prices, 1967–90
(dollars per thousand cubic feet)

Year	U.S.A. Wellhead (*a*)	Residential (*b*)	Industrial (*c*)	Interstate Pipeline Margin (*c* – *a*)	Gas Utility Margin (*b* – *c*)
1967	$0.16	$1.04	$0.34	$0.18	$0.70
1970	0.171	1.09	0.37	0.20	0.72
1975	0.445	1.71	0.96	0.52	0.75
1980	1.59	3.68	2.56	0.97	1.12
1985	2.51	6.12	3.95	1.44	2.17
1986	1.94	5.83	3.23	1.29	2.60
1987	1.67	5.54	2.94	1.27	2.60
1988	1.69	5.47	2.95	1.26	2.52
1989	1.69	5.64	2.97	1.28	2.67
1990	1.71	5.80	2.93	1.22	2.87

Source: Capital Guardian Research Co.

Note: Margins shown are intended to show only who is gaining economic rent.

the burden of giving the pipelines a satisfactory return on equity. Politicians and local public utility commissions are paying close attention to this possibility because consumers are their constituents. If FERC shifts from a producer majority to a consumer majority under the Clinton administration, Order 636 might be modified.

On the other hand, some argue that Order 636 was so difficult to produce, FERC does not want to deal with it again; so it will go forward as planned.

Valuation

Analysts use several different approaches to value local distribution companies and interstate pipeline companies based on the differing characteristics of the two industry segments. Debt claims on some pipeline companies have typically exceeded the rate base, which is why the companies need to offer equity to investors. Finding that value is where analysts spend most of their time.

Table 2 presents a summary valuation for 13 pipeline companies. On average, the ratio of total capitalization to rate base is 2.46, or about half the ratio of net debt to cash flow. In many cases, the net debt is greater than the rate base. If net debt is greater than the rate base and returns are in the single digits, the company has a problem. The result will be the relative performance Richard Gross shows in his

Figure 3. LDC versus Total Natural Gas Industry Sales, 1970–89

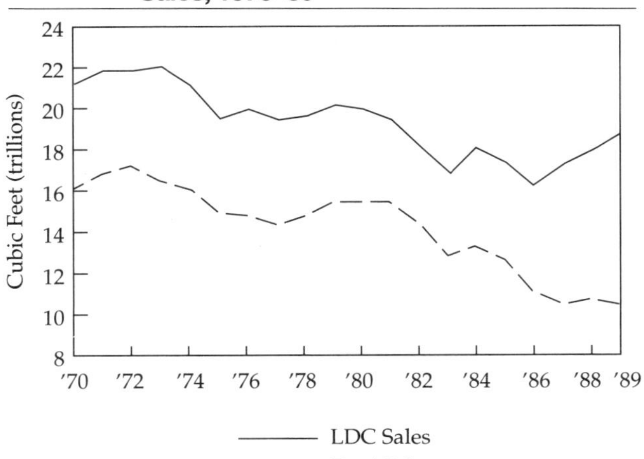

Source: Capital Guardian Research Co.

Figure 4. Wellhead Price as a Percentage of Residential, Commercial, and Industrial Prices, 1980–91

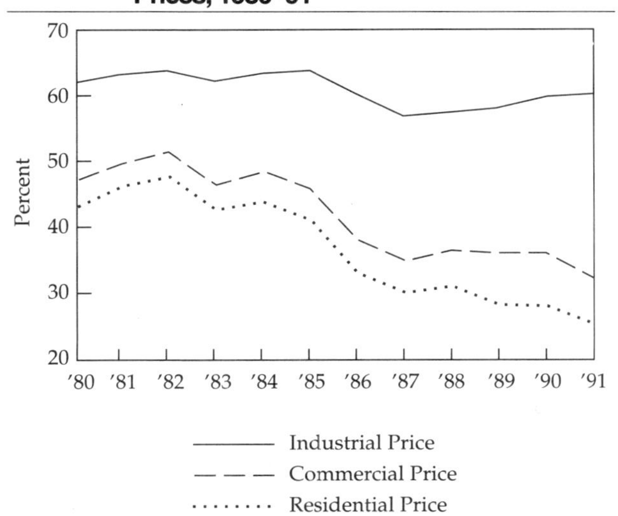

Source: Capital Guardian Research Co.

Table 2. Summary of Pipeline Stock and Financial Data

	Stock Price Data								
Company	Closing Price 11/6/92	Price 12/31/91	Percent Change	Total Capital/Rate Base	Total Capital/Total Asset Value	1993 P/E	Debt – Adjusted Market Value/Cash Flow before Interest, 1993	Yield	Projected Net Debt/1993 Cash Flow
Arkla	8.625	12.500	–31.0%	1.24	1.05	21.56	6.50	3.25%	5.50
Consolidated Natural Gas	46.250	43.000	7.6	2.37	1.68	17.13	8.51	4.06	2.32
Coastal Corp.	25.250	24.625	2.5	3.71	1.89	15.78	8.47	1.58	8.16
Enron Corp.	48.250	35.000	37.9	5.25	1.70	16.08	8.22	2.90	3.00
Equitable Resources	47.375	40.875	15.9	2.92	1.29	13.16	7.37	3.42	2.44
Panhandle Eastern	19.125	15.375	24.4	2.55	2.19	15.30	7.85	4.18	5.75
Questar Corp.	25.375	21.375	18.7	2.00	1.51	14.50	7.34	4.02	2.05
Sonat	43.000	33.000	30.3	3.37	1.06	20.48	7.47	4.65	3.69
Transco Energy	14.625	19.000	–23.0	1.51	1.24	14.63	6.43	4.10	8.40
Williams Companies	37.375	38.375	–2.6	2.68	1.08	13.84	7.37	4.07	4.93
Columbia Gas System	22.500	17.250	30.4	1.44	1.16	15.00	6.80	0.00	7.14
Pacific Enterprises	19.000	26.250	–27.6	1.14	1.14	11.52	6.29	0.00	4.55
El Paso Natural Gas	29.750	19.000	56.6	1.74	1.74	13.22	7.78	3.36	3.56
Average			10.8%	2.46	1.44	15.55	7.42	3.05%	4.73

Source: Capital Guardian Research Co.

Notes: Total capital = Market capitalization + Net debt + Preferred stock. Total asset value = Rate base + Earnings/Price reserve base + Other assets. Total capital/Rate base = Value of the company relative to the utility assets. Total capital/Total assets = Value of the company relative to all assets, including diversified businesses. Debt – Adjusted market value = Market capitalization + Net debt.

Figure 5. Allocation of the Average U.S. Residential Price, 1984–91

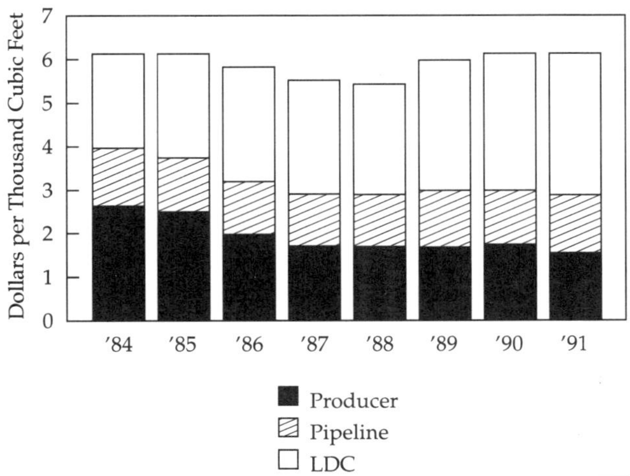

Source: Capital Guardian Research Co.

presentation of historical shareholder returns.[1]

Order 636 changed the perception of expected investor returns. Investors have historically perceived pipeline returns as highly risky but, on average, low. Under Order 636, pipeline returns should be less risky, because covering costs and profits in the demand charge should improve the companies' chances of earning the allowed rate of return. In response to Order 636, therefore, stocks have rallied from a relatively low base. Analysts can make their own judgments as to whether the stocks are expensive or cheap on an absolute basis. Clearly, on a P/E basis, some positive fundamentals are attracting buyers.

Figure 6 can help analysts establish which pipeline stocks are expensive relative to others in the industry. The figure plots price-to-book-value (P/BV) ratios against ratios of debt-adjusted market value to after-tax cash flow before interest (CFBI). Many natural gas companies trade at about 1.5 times book value, the industry quality standard. Companies such as Arkla represent good value, whereas Enron is expensive on a book-value basis.

Rate-base growth for LDCs in the 1980s was exceptional, as shown in **Figure 7**. During this period, although sales were essentially flat and allowable returns on equity declined, LDCs' rate-base increases more than made up for declines in allowed return on equity. Since peaking in the late 1980s, rate-base growth has slowed significantly, and the deceleration is expected to continue because of the group's capital expenditure programs.

For analyzing LDCs, one useful but simple valuation tool is to measure earnings per share (EPS)

Figure 6. Relative Costliness of Stocks—Diversified U.S. Pipeline Companies

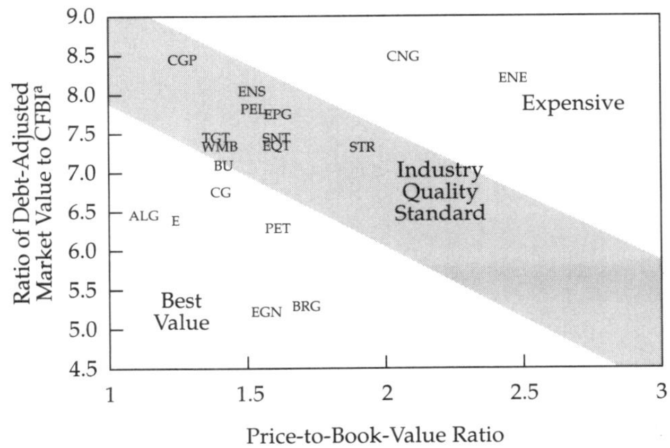

[a]After-tax cash flow before interest.

ALG	= Arkla	ENS	= ENSERCH Corp.
BRG	= British Gas	EPG	= El Paso Natural Gas
BU	= Brooklyn Union Gas Co.	EQT	= Equitable Resources
CG	= Columbia Gas System	PEL	= Panhandle Eastern
CGP	= Coastal Corp.	PET	= Pacific Enterprises
CNG	= Consolidated Natural Gas	SNT	= Sonat
E	= Transco Energy	STR	= Questar Corp.
EGN	= Energen Corp.	TGT	= Tenneco
ENE	= Enron Corp.	WMB	= Williams Companies

Source: Capital Guardian Research Co.

versus dividend payout. **Table 3** presents EPS, dividends, and the dividend payout ratio for LDCs from 1978 through 1990. In the early 1980s, the group paid out less than 60 percent of earnings in dividends. Dividends grew significantly during the decade, but earnings had a negative growth rate. By 1990, divi-

Figure 7. Average Rate-Base Additions for Composite of Natural Gas LDCs, 1981–94[a]

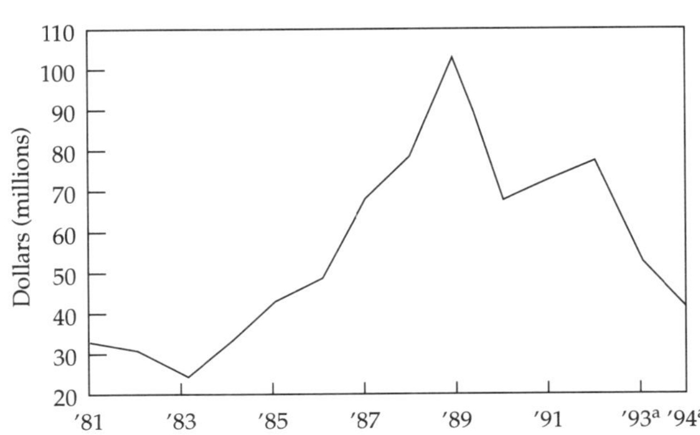

Source: Capital Guardian Research Co.

Notes: Rate-base addition is defined as capital expenditures less depreciation. The LDC composite is composed of Atlanta Gas Light, Brooklyn Union Gas, MCN Corp., National Fuel Gas, NICOR, Peoples Energy, and Washington Gas Light Co.

[a]Estimate.

[1]See Mr. Gross's presentation, pp. 130–140.

Table 3. Moody's LDC Index Earnings per Share versus Dividend, 1978–90

Year	EPS	Dividends	Dividend Payouts
1978	$7.78	$4.18	54%
1979	8.56	4.44	52
1980	8.53	4.68	55
1981	8.77	5.11	58
1982	7.46	5.39	72
1983	6.88	5.55	81
1984	7.15	5.88	82
1985	7.42	6.22	84
1986	6.93	5.71	82
1987	8.00	6.02	75
1988	9.37	6.30	67
1989	9.50	6.58	69
1990	7.62	6.84	90
1978–90 Growth rate	–0.2%	4.2%	

Source: Capital Guardian Research Co.

Figure 8. Absolute and Relative LDC Stock Price Performance, 1982–92

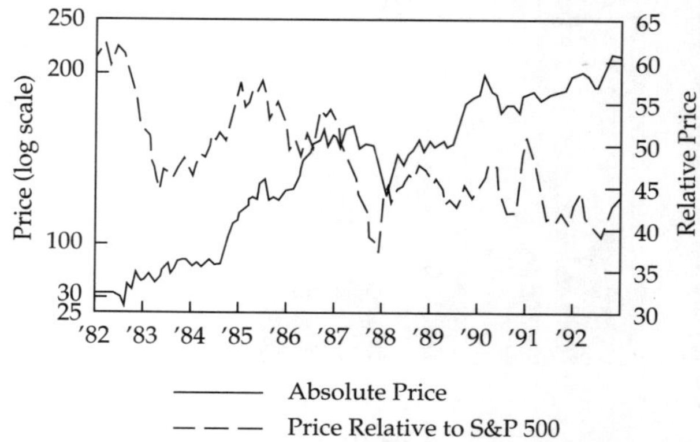

Source: Capital Guardian Research Co.

dend payouts were in the 90 percent range. Dividend growth projections used in future valuations will need to be lower than in the past.

Valuation measures relative to the market and historical ranges provide analysts with a useful perspective on current prices. Historically, only absolute returns mattered. Investors wanted to know whether they made money or not. Today, with institutional investors representing half the market, performance relative to the market has become increasingly important.

The August 1992 values and the most recent 10-year averages of various valuation measures for LDCs are shown in **Table 4**. Keep in mind that using this particular 10-year period as a benchmark may be dangerous, because it reflects the significant decline in long-term interest rates of the late 1980s. A longer term horizon that includes many different interest rate and market cycles, which would tend to balance out in an average, might be more appropriate. Note that Table 4 contains absolute and relative P/Es, P/BVs, and yields.

In August 1992, absolute P/Es and P/BVs for LDCs were at the high end of their 10-year range, and yields were at the low end. Even though market P/BV was at an all-time high, the relative values for the LDC stocks were still within their 10-year ranges.

During the past 10 years, the LDCs as a group have outperformed the S&P 500 on a total return basis by 20.1 percent to 17.5 percent, and the future performance of these stocks may be as good as the past. On a relative basis, the group is currently within its 10-year range with respect to yield, P/BV, and P/E. The LDCs also look good on the basis of a relative return on equity. These stocks are cheap if investors consider that the return on equity of corporate America has declined during this deflationary period while the allowed return on equity of LDCs has remained comparatively stable.

Figure 8 shows absolute and relative stock price performance, not including dividends, for the natural gas distribution companies. Analysts should be careful when relying on charts of relative price performance, however, because the charts typically do not consider dividend yield. Most relative price charts, because of their focus on stock price, show the LDC group underperforming the market. If Figure 8 had included dividends, it would have shown LDCs to be one of the best performing groups in the market in the past 10 years.

The yield on long-term T-bonds is strongly correlated with LDC stock prices, as shown in **Figure 9**, and this relationship can be used to signal trading points. **Figure 10** illustrates a trading mechanism that signals a sell when the relative yield of the LDC group versus the long-term T-bond yield falls below

Table 4. Various LDC Measures—August 1992 versus 10-Year Averages

	P/E	Relative P/E	P/BV	Relative P/BV	Yield	Relative Yield
August 1992	14.9	75.4%	1.6	82.0%	5.7%	193%
10-year high	16.9	129.0	1.7	86.0	14.6	250
10-year low	5.6	60.0	0.5	47.0	5.6	150

Source: Capital Guardian Research Co.

Figure 9. LDC Stock Price and the Yield on Long-Term T-Bonds, 1982–92

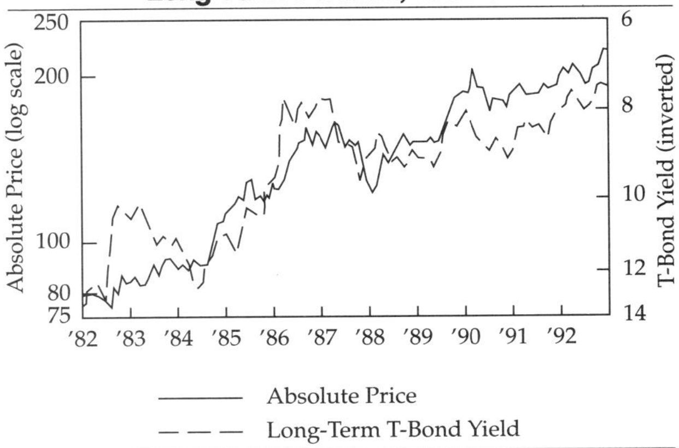

Source: Capital Guardian Research Co.

Figure 10. Using LDC Relative Dividend Yield versus Long-Term T-Bond Yield for Portfolio Decisions

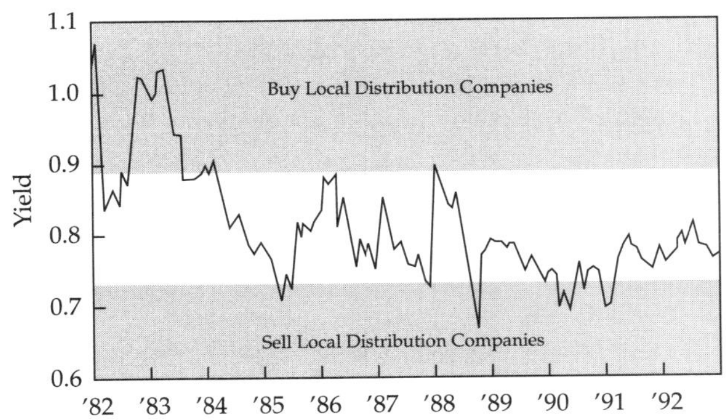

Source: Capital Guardian Research Co.

73 percent and a buy when the ratio exceeds 0.90. This figure indicates that, as of November 1992, the LDC group still has a way to go, if the long-term T-bond yield continues declining, before reaching a sell signal.

Finally, note that when analysts are following natural gas securities, they can gain a lot of information by talking to participants in industry segments other than the pipeline or distribution companies. For example, analysts who were talking only to pipeline companies would not have seen take-or-pay changes coming. The exploration and production companies, however, were more than happy to tell analysts about the change, because it supported their stock prices.

Question and Answer Session

Richard G. Gross II
Michael T. Kerr

Question: As the gas industry becomes less regulated than previously and expands its non-rate-base business, will the value of using price-to-book-value ratios diminish or increase?

Gross: The answer depends, in part, on how facilities driven the new business is. If companies decide to expand into the gathering end of the business, the new business will be facilities driven; if companies decide to generate a book of long-term swaps, the new business will not be facilities driven. The prototypes for success are already in the facilities business and they will not be spending much additional money. Most of the financing will be off the balance sheet. As a result, return on equity will improve, and investors will be more willing to pay a high price relative to book value. Successful companies will also enjoy higher P/Es.

Question: Order 636 seems to be better for performance of pipeline bonds than for pipeline equities. Do you agree?

Gross: Order 636 is great for bonds because it reduces the volatility of cash flow immensely. Equity results are mixed. For a pipeline company earning 20 percent return on equity, most of the excess return has been volume driven. Regulatory revenue collection is based on recovery costs and return on investment over a specific volume; if rates were designed to collect more than 100 units, companies moving 120 units generated an incremental return. Good companies were able to outperform the allowed return on equity by 500–700 basis points. In the new regulatory environment, the ability to earn an incremental return will be limited, and a good company might be able to earn 200–300 basis points more than the allowed rate of return. For companies that previously earned 20 percent and are now scaled back, Order 636 is a negative. Companies that were earning 8 percent because they were discounting rates can now return to normal rates; so they are better off.

Kerr: Order 636 will not bring salvation for pipeline companies that do not connect production sites with markets. Pipelines that serviced the flow of gas from the Gulf of Mexico north will lose, because the main flow is now west to east. The pipelines in the middle of the flow will earn above-average rates of return. The good pipelines will make rates of return that are better than the allowed rates, and the bad pipelines will not.

Question: What is the major issue facing pipeline companies?

Kerr: Future problems will not be anything close to what the industry has already been through. Everything that could have possibly gone wrong in the past decade did. The magnitude of the take-or-pay hits, the lack of a guaranteed market or a guaranteed return, and market pressure on margins all contributed to the hostile environment for the industry. I do not think the future holds anything of similar magnitude. Some disruption may occur when the new FERC majority takes over, and some debate will take place at local public utility commissions about whether transition costs will flow through, but from a fundamental viewpoint, the next five years will be friendlier than the past five.

Gross: The next big problem will not involve macro issues, but micro issues will arise for individual stocks. For example, companies with limited opportunities for rate-base growth will generate lots of cash flow, and their biggest challenges will be to ensure quality management decisions regarding reinvestment. The pipeline companies have not in the past shown an ability to diversify and generate very exciting rates of return.

Some companies have been damaged to the point that their balance sheets need drastic repair. The industry has some A-rated companies, some companies clinging to investment-grade ratings, some with split ratings, and some with ratings below investment grade. Their abilities to get their balance sheets back in order will reflect the quality of their management decisions.

Question: How do you adjust for negative working capital?

Gross: I make a superficial adjustment. Pipelines and producers need very little working capital: For most, payables equal receivables, plus some accrued taxes and pipeline storage, which builds and then comes back down through the years. The big imbalances for pipelines derive from off-balance-sheet financing.

They factor receivables, do not offer customer refunds with an escrow account, and use the money for other purposes. When the refunds are due, the company does not have the money. To adjust for this circumstance, I add negative working capital to debt. It is not a fair approach for companies with true working inventories, such as Coastal, but that is the subjective part of the valuation process.

Question: How will competitive pressures affect the transmission companies under Order 636? What will be the most important factors shaping future market share?

Kerr: The main competitive pressures in the next five years will be in the growth areas where LDCs have a disadvantage because of higher costs. A lot of focus will be on the bypass issue. For example, Atlanta Gas Light is Sonat's biggest customer, and for a long time, Sonat refused to connect a particular large industrial customer that wanted to bypass Atlanta Gas Light. In 1991, Sonat was ordered to complete that bypass and give the industrial customer the advantage of the cheaper gas.

Three years after Order 636 becomes law and the return on equity is earned in the demand charge, the pipelines will try to earn an above-average rate of return with their gas marketing subsidiaries. They will not have to file a purchased gas adjustment, and they will not have to go back through the rate-case process, so any additional volume they can carry on the system will contribute to an above-average rate of return. Pipelines will have an incentive to pursue growth in the gas business.

The LDCs' attitude toward natural gas marketing has improved significantly from five years ago. They use incentives to encourage gas sales forces to sign contracts and gain markets. The local utility commissions fully understand that, because of the bypass issue, an industrial customer cannot be expected to pay the full load a residential customer pays. For the most part, regulatory schemes have leveled the playing field. During the next five years, the average LDC will earn its allowed return on equity but not much more than that.

Question: How do you measure regulation as a variable in the model correlating historical performance as a function of natural gas prices, book value, regulations, and so forth?

Gross: Other than regulation, all the factors in the model are hard numbers. The regulatory impact is a subjective number, but it is easy to fit the curve historically. The correlation can be improved by tinkering with the severity of the regulatory impact number.

If you eliminate the regulation number entirely and assign a constant value of 1 for a positive environment and –1 for a negative environment, it still correlates well. For example, Order 380 was not perceived to be negative; take-or-pay was no real threat to the industry until oil prices fell. So Order 380 would not have changed the constant value of 1. The investment community considered Order 636, "Take-or-Pay Round Two," the end of the world because the balance sheets of these companies were very precarious. As a result, the impact on the model of Order 636 was much greater than that of Order 380.

Question: Should an investor play the natural gas stocks as long-term investments or as seasonal, weather-related trading vehicles?

Kerr: My firm clearly invests in these stocks for the long term. We do not want to trade them very aggressively. Instead, we have bought stocks with three- or four-year horizons. A fundamental rule of investing is that research does not bring much added value to any pure commodity investment. Virtually every investment we have made in this industry relies on something other than a rise in commodity prices to increase the investment's value.

For example, for the most part, the leverage factor went the wrong way in the 1980s. Therefore, if analysts can identify some reason the leverage of some companies will improve during the coming three- or four-year period, either through asset sales or a better return on equity, they can foresee added value to the stocks. All the companies have been in a fairly significant rate-base-building mode and will start to generate increased free cash flow, so leverage in this industry could decline significantly in the next five years.

Most years, investors will wish they had sold the stocks in October and bought them again in March, but the hope is that each year will bring higher highs to make up for the seasonal dips.

Gross: The seasonality in stock prices is attributable to commodity price seasonality—a big factor in the short-term movement of the stocks. Because the gas business is changing, however, investments based solely on seasonality will not necessarily work. During the first 11 months of 1992, the big influence on stock prices was Order 636, not seasonality; selling in the spring would have been the worst strategy. Regulatory impacts are not seasonal. Future commodity prices may not be as seasonal as they were before Order 636.

Question: Assuming that no challenge to Order 636 is successful and pipeline expansion continues, is overbuilding a possibility?

Gross: Overbuilding is not only possible, it is already occurring. California is an overbuilt market; off-peak capacity is in significant oversupply in New England; and the whole Midwest is awash in excess capacity. As Order 636 is implemented, high-cost pipelines in each of those areas will experience competition problems and margin erosion as they discount tariffs to encourage pipeline utilization.

Glossary

Allocated cost of service. Total fixed and variable costs associated with transporting natural gas. These costs are allocated on the basis of expected volume.

Barrel. A barrel of petroleum contains 42 gallons.

Barrel-of-oil equivalent (BOE). Term used to present oil and natural gas reserves on a comparable basis. One barrel of oil contains six times more Btus than a thousand cubic feet of natural gas. Also known as *oil-equivalent barrel.*

Base load. A distributor's given send-out of natural gas that remains fairly constant over a period of time and is usually not sensitive to temperature. Such gas is used in the residential market to operate clothes dryers, water heaters, and cooling systems.

bcf. 1 billion cubic feet, or 1 million thousand cubic feet; usually used as a measure of natural gas volume.

Btu (British thermal unit). A measure of the quantity of heat required to raise the temperature of 1 pound of water 1 degree Fahrenheit (58.5°F to 59.5°F) at a standard pressure of 30 inches of mercury.

Budget survey. Annual survey reflecting the anticipated capital spending of industry members.

Bypass. The process of selling natural gas directly from producer or pipeline to end-user and thus, in effect, bypassing the local distribution company. Also known as *LDC bypass.*

City gate. Point at which a gas distribution company receives natural gas from a pipeline company.

Cogeneration. A process by which otherwise wasted heat is used to produce electricity and thermal energy at an end-user site from a single energy input source.

Cost of service. In public utility regulation, the total number of dollars required to supply any total utility service (i.e., the revenue requirement), including operation and maintenance expenses; other necessary costs such as taxes, depreciation, depletion, and amortization of property not covered by ordinary maintenance; and a fair return so the utility can maintain its financial integrity, attract new capital, and compensate its owners for the risks involved.

Cut point. In the distillation process, the temperature at which a given product is extracted from the feedstock. For example, the cut point for gasoline is 200°F.

Degree day. Measure of the variation of the mean daily temperature from a reference temperature, usually 65°F. Mean temperatures below 65°F result in heating-degree days and mean temperatures above 65°F result in cooling-degree days. For example, on a day when the mean outdoor temperature is 35°F, 30 heating-degree days will be experienced.

Deliverability. The volume of natural gas a well, field, pipeline, storage reservoir, or distribution system can supply in a given period of time.

Distribution. The process of distributing natural gas from the city gate, or plant, to consumers. Also, a functional classification relating to that portion of utility plant used for delivering gas from the city gate to consumers or to expenses relating to the operation and maintenance of a distribution plant.

Distribution company, gas. A company that obtains the major portion of its gas operating revenues from the operation of a retail gas distribution system and that operates no transmission system other than incidental connections within its own system or to the system of another company. See also *Local distribution company.*

Equity production. Production by a party with at least a partial ownership interest in the oil or natural gas reserves. Because of the equity interest, the owner benefits from upward movement in crude oil and/or natural gas prices.

Estimated potential gas reserves. The total amount of natural gas estimated to exist in a specified area, whether or not it is considered proven or recoverable.

Estimated proven recoverable gas reserves. The estimated quantity of natural gas that analysis of geological and engineering data demonstrates with reasonable certainty to be recoverable from known oil and gas reservoirs under existing economic and operating conditions. Reservoirs are considered proven if they have either demonstrated the ability to produce by actual production or passed conclusive formation tests.

Federal Energy Regulatory Commission (FERC). An independent, five-member agency that regulates interstate commerce involving natural gas, electrical power, and all other forms of energy. FERC regulates producer sales of natural gas in interstate commerce and establishes uniform ceiling prices for each of several categories of natural gas that apply to all sales nationwide. It also has the added responsibility of regulating rates and tariffs of oil pipelines.

FERC Order 380. Issued May 25, 1984. Eliminated variable costs from the minimum commodity-charge portion of natural gas pipeline sales tariffs.

FERC Order 636. Issued April 6, 1992. Unbundled natural gas pipelines' transportation, storage, and other services, terminated the merchant function of pipelines, allowed secondary markets for capacity, and required electronic bulletin boards.

Feedstock. Raw material supplied to a machine or processing plant.

Firm customers. Those customers under schedules or contracts for continuous natural gas delivery over a specified period of time.

Independents. Small oil and natural gas companies not affiliated with the major oil companies.

Industrial fuel switching. Switching from natural gas to alternative fuels such as residual or clarified fuel oil; primarily motivated by the fuels' relative prices.

Interruptible service. Low-priority natural gas service offered to users under contracts that anticipate and permit interruption on short notice, generally in peak-load seasons, so as to supply firm customers and higher priority users. Unlike off-peak service, interruptible service makes gas available at any time of the year.

Liquefied natural gas (LNG). Natural gas that has been liquefied by reducing its temperature to –260°F at atmospheric pressure. It remains a liquid at –116°F and 673 PSIG. The liquid volume is 1/625 that of the gas volume at ambient temperature and pressure. Also known as *liquid natural gas* and *natural gas liquids* (NGL).

Local distribution company (LDC). Natural gas distribution company that buys from pipelines and distributes to end-users. Also known as *gas utility*.

Majors. Large integrated oil companies. See also *Seven Sisters*.

mcf. 1,000 cubic feet; usually used as a measure of natural gas volume.

mmBtu. 1 million British thermal units. There are 950–1,200 mmBtu in 1 cubic foot of natural gas. Under normal conditions, 1.030 mmBtu equals 1 mcf.

mmcf. 1 million cubic feet; usually used as a measure of natural gas volume.

Natural gas. A naturally occurring mixture of gases found, often in association with petroleum, in porous geological formations beneath the earth's surface. The largest constituent of natural gas is methane.

New gas. Natural gas made available by a contract of purchase and sale after the enactment of the Natural Gas Policy Act of 1978.

Nonequity production. Oil production performed on a contract basis. The resource owner pays the producer a fee for every barrel of oil produced. The producer under contract has no equity in the oil reserves and is insulated from fluctuations in crude oil prices.

No notice. A delivery service of natural gas pipelines that allows LDCs to draw their maximum load on demand without penalty. The pipeline must provide storage, transmission, and load balancing as part of this service. Under FERC Order 636, these services will be unbundled, and LDCs will have to rebundle them to replicate no-notice service.

Normalization. For rate-making purposes, the adjustments to sales, revenues, and expenses of a historical test year to reflect differences from expected normal weather patterns.

Old gas. Natural gas made available by a contract of purchase and sale before the enactment of the Natural Gas Policy Act of 1978.

OPEC (Organization of Petroleum Exporting Countries). Members are Algeria, Gabon, Indonesia, Iran, Iraq, Kuwait, Libya, Nigeria, Qatar, Saudi Arabia, the United Arab Emirates, and Venezuela.

Peak load. The greatest demand on a natural gas pipeline or energy distribution system during a specified period of time.

Peak shaving. Supplying fuel gas to a distribution system from an auxiliary source during periods of maximum demand when the primary source is inadequate.

Pipeline. All parts of the physical facilities through which natural gas is transported, including pipe, valves, and other appurtenances attached to the pipe, compressor units, metering stations, regulator stations, delivery stations, holders, and fabricated assemblies. See also *Transmission*.

Plowback. The percentage of wellhead revenue dedicated to reserve replacement.

Producer. The owner of wells producing oil or natural gas.

PSIG (pounds per square inch gauge). This pressure is measured relative to atmospheric pressure instead of relative to a vacuum. Thus, absolute pressure is equal to PSIG less atmospheric pressure.

Public utility commission (PUC). A state regulatory authority set up to monitor monopolies granted to intrastate and local natural gas pipelines. The primary responsibility of a state PUC is to regulate the rates that privately owned utilities involved in intrastate commerce charge their customers.

Rate base. The value established by a regulatory authority on which a utility is permitted to earn a specified rate of return. Generally, the base represents the amount of property used and the values or combinations of values of the following: fair value, prudent investment, reproduction costs, and/or original cost. Rate base may include cash, working capital, materials and supplies, and deductions for accumulated provision for depreciation, contributions in aid of construction, accumulated deferred income taxes, and accumulated deferred investment tax credits.

Rate case. The process through which regulated natural gas billing rates are negotiated.

Rate design. The method of classifying fixed and variable costs between the demand and commodity components of a natural gas company's rates.

Rate of return (ROR). The return earned or allowed to be earned by a utility; calculated as a percentage of its fair value or rate base.

Reserve replacement ratio. The ratio of new or additional reserves to reserves produced or developed in a given year.

Seven Sisters. Originally, British Petroleum, Chevron, Exxon, Gulf Oil, Mobil, Royal Dutch/Shell, and Texaco. (Chevron acquired Gulf Oil.)

Spot market. Short-term contract sales of natural gas, crude oil, refined petroleum products, or liquid gas.

Take-or-pay. The clause in a natural gas supply contract providing that, for a specific time period, a purchaser must pay for a specified minimum quantity of gas regardless of whether the purchaser accepts delivery. Some contracts allow the buyer to take later delivery without penalty.

tcf. 1 trillion cubic feet; usually used as a measure of natural gas volume.

Test year. The 12-month period selected as the base for presenting data in a case or hearing before a regulatory agency.

Theoretical crack spreads. The theoretical spread or difference between the cost of feedstock and the value of the products it yields. The calculation is based on cracking one barrel of crude oil into one barrel of product. A ratio defines the output based on the particular type of refinery. For example, the 3:2:1 crack spread is an industry benchmark indicating 3 parts of oil will be processed into 2 parts gasoline and 1 part distillate. To determine the theoretical crack spread, an analyst calculates the weighted-average value of the final products and then subtracts feedstock costs. Analysts consider the spread to be theoretical because actual production may vary from the hypothetical output.

Transmission system. Pipelines installed for the purpose of transmitting natural gas from sources of supply to distributing centers or large-volume customers. Transmission lines operate at higher pressure, are longer, and have a greater distance between connections than distribution lines. Also known as *transportation system.*

West Texas Intermediate (WTI). Crude oil used as a benchmark for oil prices.

Order Form022

Additional copies of *The Oil and Gas Industries* (and other AIMR publications listed on page 155) are available for purchase. The price is **$20 each in U.S. dollars**. Simply complete this form and return it via mail or fax to:

AIMR
Publications Sales Department
P.O. Box 7947
Charlottesville, Va. 22906
U.S.A.
Telephone: 804/980-3647
Fax: 804/977-0350

Name ______________________________

Company ______________________________

Address ______________________________

______________________________ Suite/Floor ______________

City ______________________________

State ______________ ZIP __________ Country ______________

Daytime Telephone ______________________________

Title of Publication	Price	Qty.	Total

Shipping/Handling

- ❑ All U.S. orders: Included in price of book
- ❑ Airmail, Canada and Mexico: $5 per book
- ❑ Surface mail, Canada and Mexico: $3 per book
- ❑ Airmail, all other countries: $8 per book
- ❑ Surface mail, all other countries: $6 per book

Discounts

- ❑ Students, professors, university libraries: 25%
- ❑ CFA candidates (ID #__________): 25%
- ❑ Retired members (ID #__________): 25%
- ❑ Volume orders (50+ books of same title): 40%

Discount $-________

4.5% sales tax
(Virginia residents) $________

8.25% sales tax
(New York residents) $________

7% GST
(Canada residents, #124134602) $________

Shipping/handling $________

Total cost of order $________

❑ Check or money order enclosed payable to **AIMR** ❑ Bill me

Charge to: ❑ VISA ❑ MASTERCARD ❑ AMERICAN EXPRESS

Card Number: ______________________ ❑ Corporate ❑ Personal

Signature: ______________________ Expiration date: __________

Selected AIMR Publications*

Execution Techniques, True Trading Costs, and the Microstructure of Markets, 1993 . . . . . $20
Katrina F. Sherrerd, CFA, *Editor*

Investment Counsel for Private Clients, 1993 . . . . . $20
John W. Peavy III, CFA, *Editor*

Active Currency Management, 1993 . . . . . $20
Murali Ramaswami

The Retail Industry—General Merchandisers and Discounters, Specialty Merchandisers, Apparel Specialty, and Food/Drug Retailers, 1993 . . . . . $20
Charles A. Ingene, *Editor*

Equity Trading Costs, 1993 . . . . . $20
Hans R. Stoll

Options and Futures: A Tutorial, 1992 . . . . . $20
Roger G. Clarke

Improving the Investment Decision Process—Better Use of Economic Inputs in Securities Analysis and Portfolio Management, 1992 . . . . . $20
H. Kent Baker, CFA, *Editor*

Ethics, Fairness, Efficiency, and Financial Markets, 1992 . . . . . $20
Hersh Shefrin and Meir Statman

Investing Worldwide, 1992, 1991, 1990 . . . . . $20 each

The Financial Services Industry—Banks, Thrifts, Insurance Companies, and Securities Firms, 1992 . . . . . $20
Alfred C. Morley, CFA, *Editor*

Managing Asset/Liability Portfolios, 1992 . . . . . $20
Eliot P. Williams, CFA, *Editor*

Investing for the Long Term, 1992 . . . . . $20

A New Method for Valuing Treasury Bond Futures Options, 1992 . . . . . $20
Ehud I. Ronn and Robert R. Bliss, Jr.

Ethics in the Investment Profession: A Survey, 1992 . . . . . $20
E. Theodore Veit, CFA, and Michael R. Murphy, CFA

The Transportation Industry—Airlines, Trucking, and Railroads, 1992 . . . . . $20
David G. Smith, *Editor*

Earnings Forecasts and Share Price Reversals, 1992 . . . . . $20
Werner F.M. De Bondt

*A full catalog of publications is available from AIMR, P.O. Box 7947, Charlottesville, Va. 22906; 804/980-3647; fax 804/977-0350.